S0-BWX-162

Expert System Technology

Development and Application

YOURDON PRESS COMPUTING SERIES
Ed Yourdon, *Advisor*

BENTON AND WEEKES *Program It Right: A Structured Method in BASIC*
BLOCK *The Politics of Projects*
BODDIE *Crunch Mode: Building Effective Systems on a Tight Schedule*
BRILL *Building Controls Into Structured Systems*
BRILL *Techniques of EDP Project Management: A Book of Readings*
CONSTANTINE AND YOURDON *Structured Design: Fundamentals of a Discipline of Computer Program and Systems Design*
DE MARCO *Concise Notes on Software Engineering*
DE MARCO *Controlling Software Projects: Management, Measurement, and Estimates*
DE MARCO *Structured Analysis and System Specification*
DICKINSON *Developing Structured Systems: A Methodology Using Structured Techniques*
FLAVIN *Fundamental Concepts in Information Modeling*
HANSEN *Up and Running: A Case Study of Successful Systems Development*
KELLER *Expert System Technology: Development and Application*
KELLER *The Practice of Structured Analysis: Exploding Myths*
KING *Current Practices in Software Development: A Guide to Successful Systems*
KRIEGER, POPPER, RIPPS, AND RADCLIFFE *Structured Micro-Processor Programming*
MACDONALD *Intuition to Implementation: Communicating About Systems Toward a Language of Structure in Data Processing System Development*
MC MENAMIN AND PALMER *Essential Systems Analysis*
ORR *Structured Systems Development*
PAGE-JONES *The Practical Guide to Structured Systems Design*
PETERS *Software Design: Methods and Techniques*
ROESKE *The Data Factory: Data Center Operations and Systems Development*
SEMPREVIO *Teams in Information Systems Development*
THOMSETT *People and Project Management*
WARD *Systems Development Without Pain: A User's Guide to Modeling Organizational Patterns*
WARD AND MELLOR *Structured Development for Real-Time Systems, Volumes I, II, and III*
WEAVER *Using the Structured Techniques: A Case Study*
WEINBERG *Structured Analysis*
WELLS *A Structured Approach to Building Programs: BASIC*
WELLS *A Structured Approach to Building Programs: COBOL*
WELLS *A Structured Approach to Building Programs: Pascal*
YOURDON *Classics in Software Engineering*
YOURDON *Coming of Age in the Land of Computers*
YOURDON *Design of On-Line Computer Systems*
YOURDON, LISTER, GANE, AND SARSON *Learning to Program in Structured Cobol, Parts 1 and 2*
YOURDON *Managing Structured Techniques, 3/E*
YOURDON *Managing the System Life Cycle*
YOURDON *Structured Walkthroughs, 2/E*
YOURDON *Techniques of Program Structure and Design*
YOURDON *Writing of the Revolution: Selected Readings on Software Engineering*
ZAHN *C Notes: A Guide to the C Programming*

Expert System Technology

Development and Application

Robert Keller

YOURDON PRESS
A PRENTICE-HALL COMPANY
ENGLEWOOD CLIFFS, NJ 07632

Library of Congress Cataloging-in-Publication Data

Keller, Robert, 1939-
Expert system technology.

(Yourdon Press computing series)
Bibliography: p.
Includes index.
1. Expert systems (Computer science)
I. Title. II. Series.
QA76.76.E95K43 1987 006.3'3 86-24742
ISBN 0-13-295577-6

Editorial/production supervision and
interior design: ***Gerry Madigan and Patrick Walsh***
Cover design: ***Ben Santora***
Manufacturing buyer: ***Ed O'Dougherty***

Printed in the United States of America

10 9 8 7 6 5 4 3 2

ISBN 0-13-295577-6 025

Prentice-Hall International (UK) Limited, *London*
Prentice-Hall of Australia Pty. Limited, *Sydney*
Prentice-Hall Canada Inc., *Toronto*
Prentice-Hall Hispanoamericana, S.A., *Mexico*
Prentice-Hall of India Private Limited, *New Delhi*
Prentice-Hall of Japan, Inc., *Tokyo*
Prentice-Hall of Southeast Asia Pte. Ltd., *Singapore*
Editora Prentice-Hall do Brasil, Ltda., *Rio de Janeiro*

Contents

Foreword

Traditional business wisdom says that any good idea for a software product should be evaluated in terms of its potential return on investment before starting development of the product. A market study that analyzes potential benefits and a product plan that details development costs form the key components of a business plan that can then be analyzed and a decision made to go ahead or not.

That was the prevailing wisdom, at least, until AI burst onto the scene. For AI systems, people act as if the critical decisions are not what applications and/or products to build, but what kind of LISP machine and/or expert system shell to buy. I have heard AI systems developers argue that good hackers and good tools are essential, but good planning is not. Some justify this position by arguing that AI projects are necessarily open-ended and do not lend themselves well to specification and planning. I have heard product and program managers state publicly that AI development efforts simply cannot be structured to fit traditional systems development methodologies.

The notion that AI systems development involves more art than methodology is one of the major barriers to successful commercialization of this new and exciting technology. AI researchers have developed a number of significant, new software development techniques that will likely lead to revolutionary changes in our society. These techniques do not, however, obviate the need for a rational approach to the development of computer software.

In this book, Bob Keller shows how AI technology can fit under the umbrella of existing systems development methodology. He is uniquely qualified for the task, having published an earlier book on systems development methodology, having served in an executive capacity with one of the first companies to enter the commercial AI field, and having extensive experience as a respected consultant in the field of AI.

The sooner this perspective is widely adopted, the sooner commercially viable AI products will emerge, and the sooner AI technology will receive wide commercial acceptance. Only then can this technology fulfill its revolutionary potential.

Steven P. Shwartz
Intelligent Business Systems
Orange, CT

Introduction

What do people mean when they say "artificial intelligence"? What is an "expert system"? And why would anyone want an expert system when perfectly good human experts are available? Suppose you wanted some AI—what would you do to get it? Providing answers to these and related questions is the mission of this book.

Since long before the industrial revolution, we have sought to enhance our limited abilities by inventing mechanical contrivances: the pick and shovel, the wheelbarrow, and similar simple tools are examples that date back to antiquity. More recently motor vehicles, manufacturing robots, and, finally, computers have been added to the panoply of artificial assistants.

It should come as no surprise, then, that technology continues to move forward, enabling us to build tools that further enhance our limited, though uniquely human, capabilities. Until recently, most of these productivity tools were developed as aids to performing brute labor or clerical tasks. In fact, the computer has become widely accepted as an assistant that is exceptionally competent at well-defined, repetitive tasks. Now researchers have developed an entirely new kind of tool: a computer that serves as an assistant whose skills include finding reasonable solutions to problems for which there may be no hard and fast "right" answers. Rather the "expert" computer system uses extensive experience-based knowledge of a subject to guess intelligently in the same way as a human expert.

1. Some Examples of Expert Systems

Imagine a machine that can listen to your medical problems with the ears of an expert diagnostician and even prescribe therapies to alleviate your dis-ease. Would you want to replace your personal physician with a machine? Of course not—drugs, appliances, and other technological devices are merely aids in the healing process, which most importantly requires interaction between patient and physician.

Nonetheless, physicians spend more and more time keeping up with the latest advances in medical technology and less and less time understanding the patient as a kindred human being. Systems such as the automated expert diagnostician could free the physician to spend more time relating to you as a human patient rather than fiddling with medical technology.

In another scenario, imagine yourself as an administrative assistant who reviews dozens of expense reports for accuracy and legitimacy. Suppose you had a computer system which could look over expense reports and spot the frauds in the same way you do. Your time might be freed to look at such interesting things as improving the company's expense policies.

Finally, imagine yourself having more bills to pay one month than you have money to spend. Short of having an automated money generator, it would be nice, time-saving, and financially effective to have on your home computer a system with the expertise of a financial planner. It could look at potential future income and expenses and recommend which current obligations are most important to pay. By analyzing your total financial situation, it might even suggest short- and long-term strategies for improving your financial condition.

2. Where Are We?

Computer technology now exists that allows us to do these things and a wide variety of other professional-level tasks that most people would agree require something we call "intelligence." Stated simply, artificial intelligence means programming a computer to perform tasks that require human-like intelligence.

These computerized tasks include communicating with you in conversational natural language, assisting you in situations that require complex decision making or planning, and, for robots, exhibiting a human-like understanding of their physical environment. Many such systems have been built to date, and many more are in progress.

Everywhere I go as a consultant or lecturer, I hear the underlying question, however subtly, about whether the emergence of AI means that computers are taking over. The answer is absolutely no. The present attempts of AI to "clone" human intelligence are at best primitive, and there is no evidence that it is even possible to emulate real human intelligence on today's computers. Just as the pick and shovel, in their narrow application areas, have not obsoleted the human hand, so the focused and narrow knowledge-based expert systems are not likely to obsolete the human brain.

3. A New Age

We are standing at the dawn of a new age, a renaissance in the computer and life sciences where people and machines work together not as master and slave, but as intelligent partners. It is an age that requires vision and understanding of the future we wish to build, an age in which metaphysical questions about the meaning of life become legitimate, and an age in which intuition, creativity, and other "uniquely" human virtues are laid open to close study.

This new renaissance holds not only vast opportunities to improve our lives and make fortunes, but also a responsibility for every individual to seek an awareness of what it truly means to be human. Here at last is a viable opportunity to begin to heal the long-standing schism between science and spirituality.

4. My Basic Attitudes

AI is here! But it's not here in all ways, or for everybody in equal measure.

My attitude in this book is that for those few companies who have plenty of money for long-term research, large knowledge-based system projects may now make sense. But for most companies that are just beginning to look at AI, undertaking a multi-million dollar knowledge-based system project is the wrong thing to do; even if a few of the current handful of experienced AI people could be collared to do the work, it's much too expensive and risky to make good business sense. Many traditional applications, however, can be substantially improved through the use of simple AI techniques in a controlled way. Thus, substantial short-term payoffs can be achieved at relatively low cost while engendering an atmosphere of learning and exploration for large-scale projects in the future.

Don't be intimidated by Artificial Intelligence! This book is a guide to the software engineering of knowledge-based systems in the day-to-day data processing environments found in most businesses; it emphasizes the cost-effective use of commercial AI technology in traditional system environments. It describes a system development methodology based on the now-widespread techniques of structured analysis, and discusses the ways in which traditional structured design and programming may be relevant. In addition, it suggests practical applications for several types of available commercial AI software as well as new computer hardware and programming languages.

5. What is Artificial Intelligence?

Artificial intelligence research is often defined as the search for general computational models of human intelligence. Unfortunately, nobody really knows what intelligence is, or how to be sure when someone is behaving intelligently. The definition and measurement of this elusive concept has been the subject of countless essays over centuries of philosophical thought. Many people have created psychological inventories in search of comparative measures of intelligence; others are satisfied with a more subjective consensual agreement that some person, or some act, is evidence of intelligence.

Part of the ideal purpose of this book was to give the reader a clear definition of AI, but what's being called AI today is undefinable and in fact does not exist as a unique entity. AI appears to be a new and unique application of tools from many older disciplines including operations research, cognitive psychology, and system design.

The basic research in artificial intelligence concerns itself with such areas of intelligent behavior as language understanding, reasoning, learning, and many others. These researches are not limited to the study of human intelligence—they also delve into the kinds of awareness exhibited by other animate creatures as well as plants.

The primary goal of AI is the machine emulation and enhancement of any and all human behaviors that have not already been automated. The pattern seems to be that once an AI problem is solved—natural English database query for example—it is no longer considered to be AI. Thus the term *artificial intelligence* really refers to an ever-evolving class of problems that AI researchers consider worthy of solution. By the time these solutions reach commercial maturity we are left essentially with some new tricks to add to our bag of data processing magic.

6. Focus of the Book

In this book the focus is predominantly on the business applications of what is currently one of the most commercially viable sectors of AI: **knowledge-based systems** and the subset of these that are called **expert systems.** Frequently the terms are used interchangeably, although to be absolutely correct, a knowledge-based system can't truly be called an expert system until its knowledge base has matured to expert status. Of equal importance, I am concerned about integrating this new technology into existing data processing environments, and so I have updated and emphasized the use of structured analysis techniques in developing knowledge-based systems.

This focus on business applications is not intended to diminish the relevance of AI to manufacturing and other areas, but simply to fill some rather serious gaps in the current literature regarding the engineering of AI software in traditional data processing environments. In spite of the business focus, much of the discussion of the project life cycle, application selection, prototyping, and others is relevant to areas other than business-oriented expert systems. Furthermore, the term *AI* should be taken to mean AI in general, but particularly as it relates to knowledge-based expert systems.

As various new techniques are discussed, you may ask yourself how these techniques differ from what DP is doing now and has been doing for almost three decades. In many cases, the differences are not immediately evident—for example, representing expert knowledge as a set of IF-THEN rules looks, at base, very much like traditional programming or decision tables.

It is true that AI researchers often use well-known DP concepts and extend them in what may appear to be simple ways. Keep in mind, however, that many of the greatest discoveries are the simplest. After such discoveries are made known, one often wonders why it was never thought of before.

Throughout the book I try to describe knowledge-based system technology in ways that show how it can be integrated with existing DP techniques, and try to relate new jargon to more familiar terminology. However, what may sometimes appear to be a trivializing of AI jargon by relating it to more familiar words and concepts is in no way intended to trivialize the work of the researchers who developed these techniques. Their contributions are monumental; their insight and perseverance in discovering elegantly simple solutions to what used to be unsolvable problems should be commended.

7. Intended Audience

Expert System Technology is intended for management and professional people in all areas of business and industry; the technology of knowledge-based systems is equally applicable to the office, the factory, or to any other area where people make decisions and plans. This book is intended to make the emerging artificial intelligence industry understandable and accessible both to non-technical business people and to technical people who have little previous exposure to AI. It brings together many dangling threads of knowledge-based system information and usage into a unified look at a new industry and its applicability to people in many areas, particularly with regard to expert system technology and natural language understanding.

For the manager, we take a detailed look at issues concerning the system development life cycle as well as selecting good applications for knowledge-based system technology (Chapters 1 through 4). A case study (Chapter 5) is used to track AI system development from application selection through analysis and prototyping. In doing this we study not only the life cycle per se, but also new options and responsibilities in programming languages, hardware, and training. An underlying message is that we must be able, fairly easily, to integrate the new techniques of knowledge-based system development with existing data processing disciplines; otherwise, AI is going to have a very short, and very uninteresting, commercial career.

For the technical professional, there is more detailed discussion of the application of expert system shells as well as descriptions of what goes on inside expert systems (Chapters 6 through 9). In particular we discuss expert systems from the point of view of the representation of knowledge in a computer and the kinds of programs that utilize that knowledge. We consider also the vicissitudes of building a knowledge base, the repository that contains the system's real expertise.

Taken together, these pages offer the manager and the professional a tool kit both for beginning a corporate training and implementation strategy for AI and for beginning to work with knowledge-based systems. As with any new technology, the road to achieving its full potential is long, and the pitfalls are many. My hope is that this book will be a guide to the commercial pioneers in AI and their successors in painlessly bringing this rich technology to a truly useful state.

Chapter 1
AI System Development Methodology

The phrase "expert systems" is on the minds of many people these days. There is a feeling that by applying expert system technology to their businesses, companies can reap enormous harvests in terms of profitability, competitive edge, and cost reduction. In fact, this can be true—some of the companies that leaped into the arena early have realized large payoffs from their investments, and the same potential exists for most other companies. The problem is planning for and managing the sensible introduction of expert systems into your company.

In a certain sense, AI is a research discipline only, the fruits of which are frequently no longer considered to be AI when they have matured to a commercially usable state. What today is being called AI is not a thing unto itself which will provide salvation for any company. Rather it is a wide variety of research efforts in robotics, natural language understanding, emulating an expert's performance, and a host of other areas. These efforts have yielded some remarkable success stories over the last few years, and engendered widespread hopes of effective computer solutions to business and manufacturing problems that have stymied data processing for a long time.

The basis of these successes amounts to a fairly small set of new tools which are understood well enough that they can be trusted outside a laboratory setting. In fact, when used in the right combination with existing systems and data processing techniques, the new tools can provide significant benefits.

This chapter introduces many of the issues surrounding the evolution of a corporate strategy for integrating the expert system branch of AI and proposes the use of a well-known methodology for developing AI systems: structured system development.

1. Why a Methodology?

Many early commercial users of AI were large corporations with sizable research budgets which allowed them to dive into a highly speculative technology without fear of disastrous downside consequences. They worked as pioneers who, with few precedents to guide them, spent large amounts of time floundering about looking for the right approach, and some efforts paid off. Today, companies of all sizes are enchanted by reported benefits of successful AI systems and are anxious to get into the fray. Because of the groundwork laid by the commercial pioneers, companies are now in a much better position to work out a corporate strategy regarding AI and to apply well-known management and system development methodologies in building systems that make use of artificial intelligence.

Regrettably, however, many companies I come upon are still taking the pioneer approach to AI. This approach is described succinctly in the words of a respected colleague of mine:[1] "The approach most corporations have taken to AI is to buy a bunch of LISP machines (and maybe expert system tools), hire a stable of programmers with some AI training or ship off their own people to be trained, and let them hack."

At this time there is little justification for an AI project to be almost totally unmanaged playtime for computer hackers. Such a scattered approach is not only wasteful of scarce resources, but is also giving undeservedly bad press to AI because of low productivity. Managing artificial intelligence is very much like managing any other system project—an AI project by any other name is still a system development project. It needs the same considerations of specification, budget, and timing as does any other system project.

Structured system development techniques offer a more appropriate approach to AI system development. These techniques are now practiced widely in data processing shops and lend themselves readily to the specification and implementation of AI systems. Every AI system project consists of pieces which can be done using traditional DP techniques (the DP component) and pieces which require AI in some form (the AI component). The structured methodology helps us to separate the AI and DP components of a project, allowing each component to be developed in a way that relates the two sensibly.

2. The AI Project Life Cycle

In this section we explore the idea of a knowledge-based system, which is the subject of this book, and contrast the AI project life cycle with that of the traditional structured

[1] Dr. Steven Shwartz is Vice-President of Technology for Intelligent Business Systems of New Haven, Conn., which develops and markets AI-based packaged solutions to business problems.

project. In the following sections we shall look more closely at the phases of the AI project.

2.1 What is a Knowledge-Based System?

In general the goal of AI is to have a machine accurately emulate intelligent human activities. Some people believe we will eventually be able to build a machine which is an accurate replica of a human being, indistinguishable from the real thing. In the context of today's technology this means having the computer do intelligent things such as understanding natural English; building robots which perform activities the way a human does; and having a computer make decisions and plan as skillfully as a human.

The human performance of all these tasks involves the skillful use of large amounts of experience from daily living, and learning from those experiences often occurs in very unconscious ways. We learn to speak English by communicating with others. We learn to do tasks by trying them, sometimes succeeding and sometimes failing. In many cases the knowledge we gain from these experiences is not firm, but rather is in the form of rules of thumb which we store away someplace for use when they are relevant to a new situation.

These rules of thumb are called **heuristics** in AI jargon and they are the mainstay of the **knowledge** we try to store for use by natural English systems, expert systems, and robots. The process of getting such heuristics from a person and storing them in the computer is called **knowledge acquisition.** It is without question the most difficult and time-consuming part of any knowledge-based system project. The reason for this becomes clear when we consider that the knowledge of an intelligent human being is indeed the sum-total of that person's entire life experience. Thus, in order to have a machine accurately and completely emulate any given person, it would have to have *all* of that person's life experiences.

For now, knowledge-based systems are much less ambitious. In fact, in an expert system we typically confine our knowledge acquisition activities to very narrow **domains of expertise.** Although much larger knowledge-based systems are being built, we will look for applications in which the expertise can be represented using from a few hundred to a few thousand heuristic rules.

This is not really as restrictive as it may seem. I am reminded of the definition of "expert" I learned in the fourth grade: "Experts are people who learn more and more about less and less until eventually they know everything about nothing." Thus in the real world it is often possible to codify true expertise in a narrow domain using surprisingly few heuristics.

In any case it is this type of work we are trying to accomplish in our knowledge-based systems: intelligent human processing, much of which is based on anecdotal information and stored as heuristic rules.

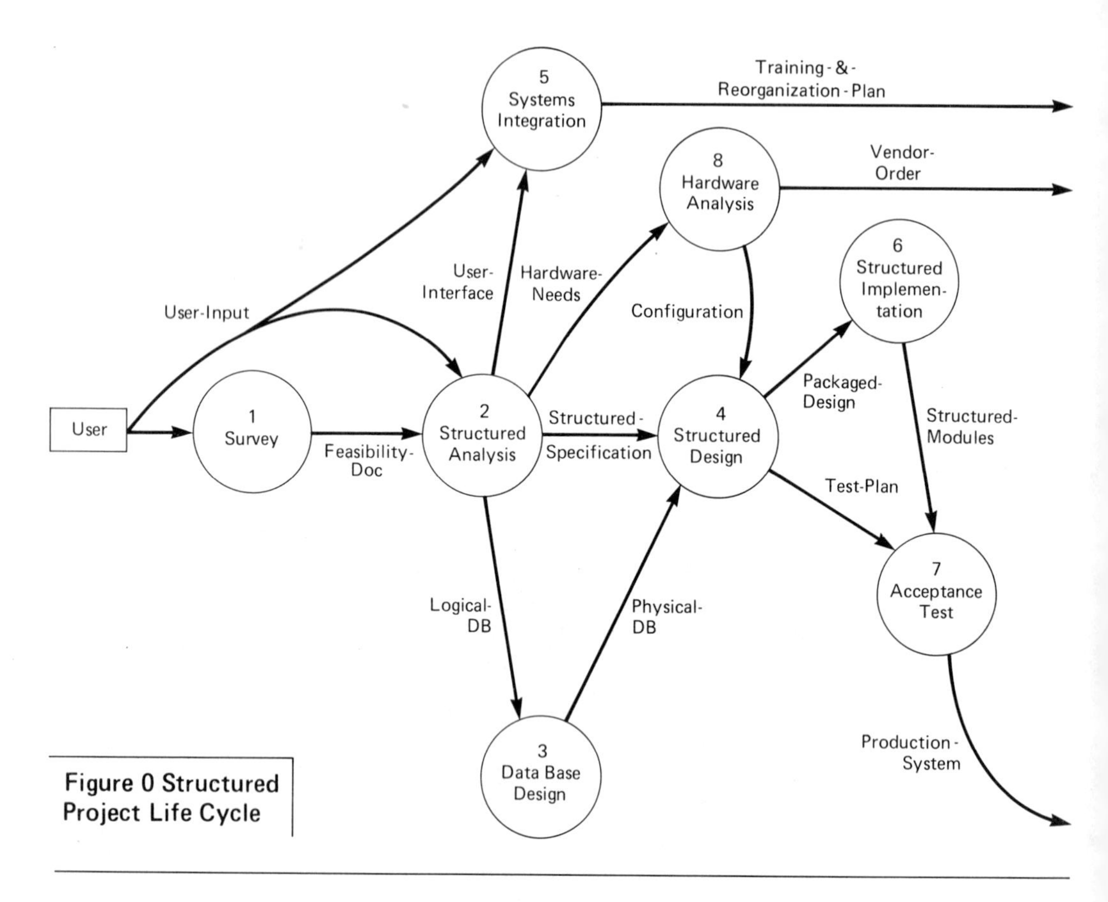

Figure 1-1. DFD Figure O. Structured Project Life Cycle

2.2 The Structured Life Cycle

A **life cycle diagram** shows the major pieces of work that go into developing a good idea into a system that usefully serves the needs of end-users in a production environment. I present three such diagrams here: the traditional structured life cycle (Figure 1-1), what appears to be the predominant current AI life cycle (Figure 1-2), and what I believe is the best life cycle for projects which include some knowledge-based processing (Figure 1-3).

Figure 1-1 shows the traditional structured project life cycle as it appears in my book *The Practice of Structured Analysis.*[2] It shows a project proceeding through four separately funded, though interrelated, activities: survey, structured analysis, structured design, and

2 Yourdon Press, New York, 1983.

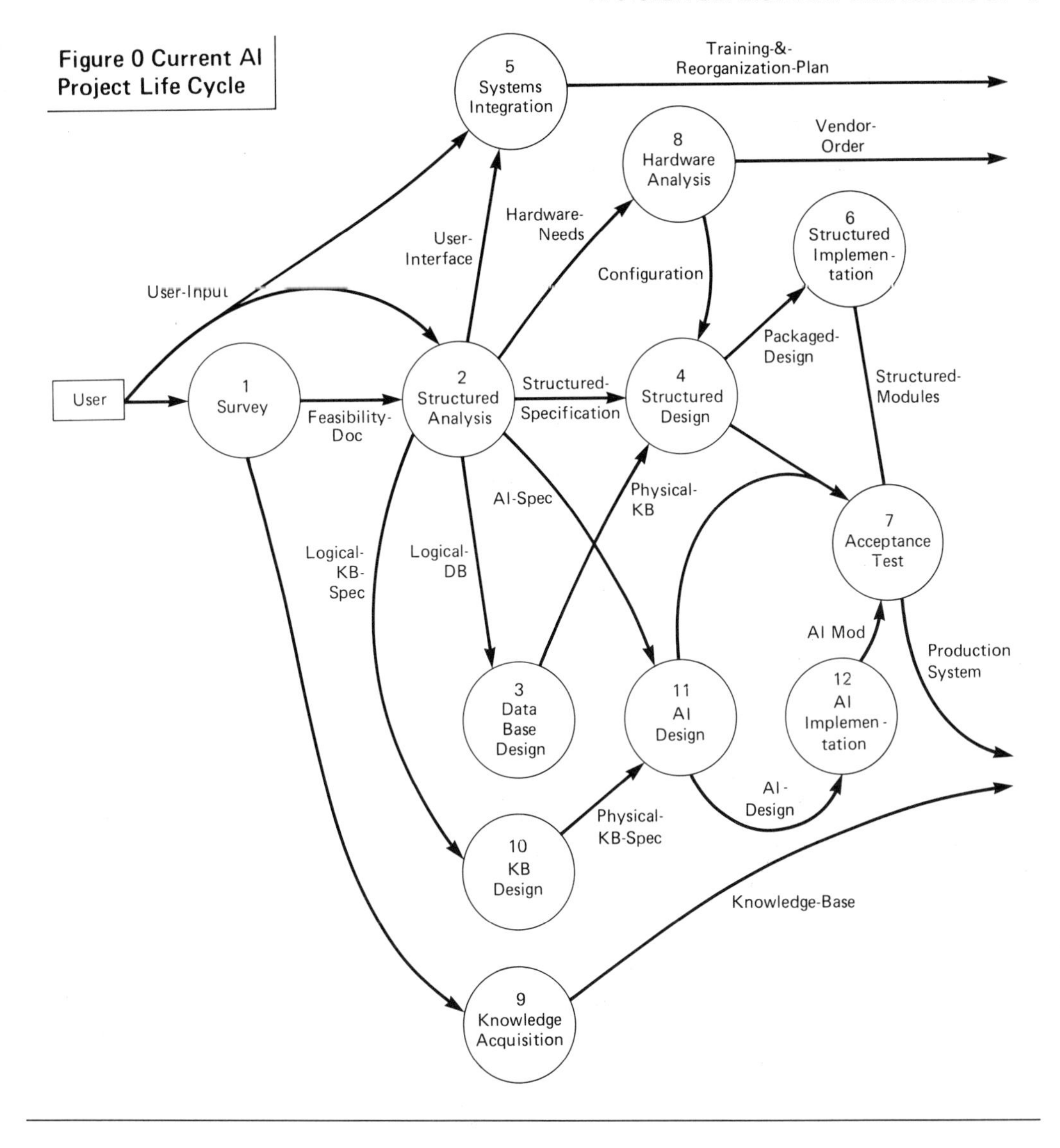

Figure 1-2. DFD Figure 0. Current AI Project Life Cycle

structured implementation. It also shows the relationship among them. Each stage requires specific kinds of work and specific documents that are to be produced.

In practice, in a business setting, structured projects tend to be oriented toward automating well-understood business tasks to produce outputs that can be supported by well-defined collections of facts called databases. The structured analysis activity results in a network diagram with narrative specifications that represent the well-defined parts of a user's needs. The structured design phase produces hierarchical charts which gener-

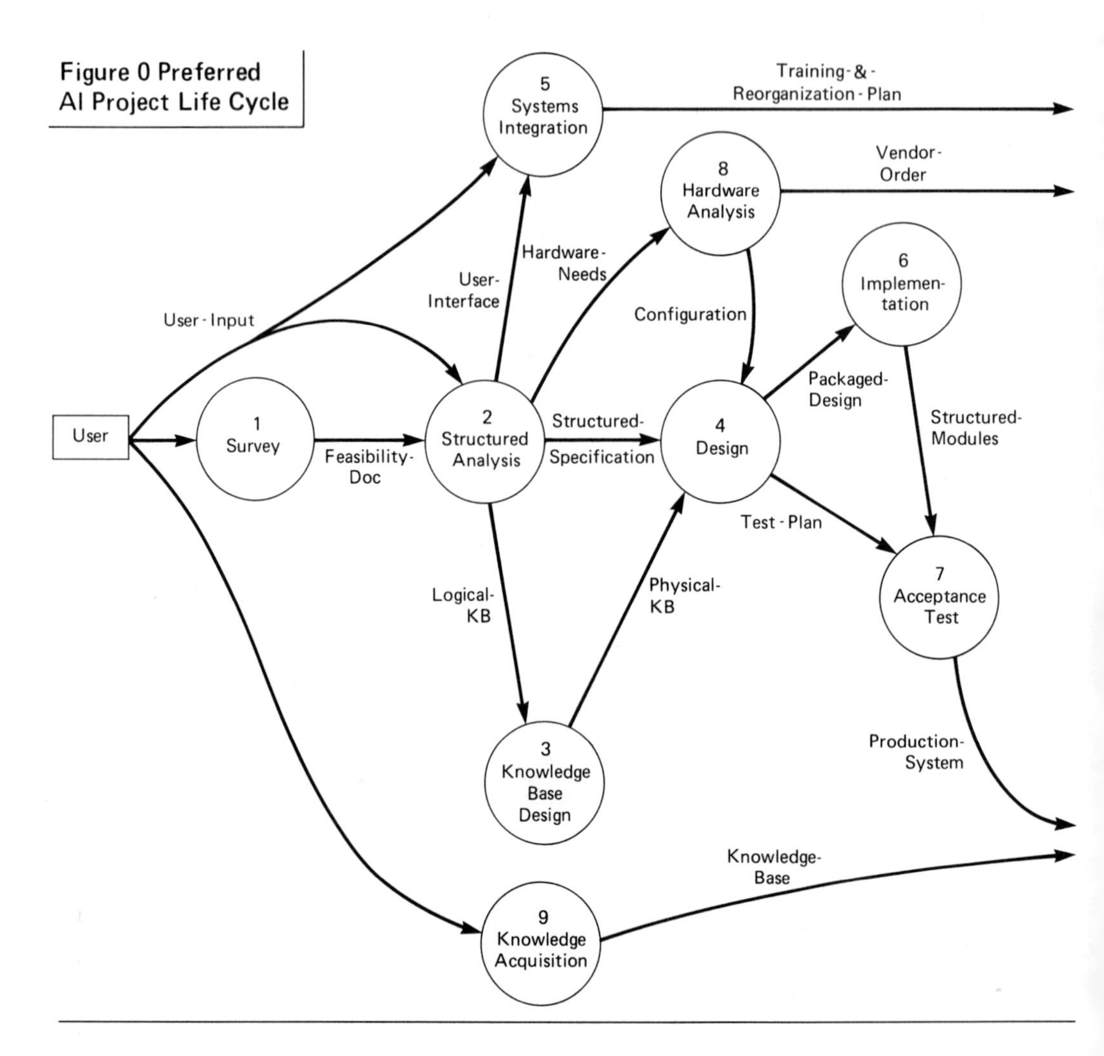

Figure 1-3. DFD Figure O. Preferred AI Project Life Cycle

ally show well-defined functional program modules being invoked by well-defined control programs with well-defined calling sequences. In structured implementation, the typical structured project will produce structured programs written in COBOL, PL/I, or some other common programming language.

To many people, the requirements for well-defined elements suggest that structured techniques are not suitable for system development that involves knowledge-based processing. In fact, the structured disciplines have quite a rich and flexible theoretical base. Unfortunately, however theoretically free of such constraints it may be, after years of use with a very restricted technology, the structured life cycle has been so constrained by its banal application that many people can't imagine using it for anything as sophisticated as AI. For these practical reasons, rather than any theoretical ones, this traditional life cycle needs some enhancement to deal with knowledge-based systems.

In too many cases, AI project management is almost non-existent, and so to try to depict the current AI project life cycle seems a bit presumptuous. Nonetheless, in those cases where there is some semblance of project management, the present situation can be approximately represented as in Figure 1-2. In this we see an attempt to keep the traditional structured approach for those parts of the project which are traditional data processing, and a parallel set of activities for the AI parts. The structured analysis carries over almost directly from the old to the new. The widespread belief that the structured approach is inappropriate for AI is evidenced by the introduction of the parallel set of development activities to deal with the AI parts of a problem.

In most cases, this is an unnecessary duplication of effort both administratively and technically. I believe that our goal should be a diagram such as that shown as Figure 1-3, which looks much more like the traditional structured life cycle. In this diagram the people, tools, and techniques of AI are integrated into a company's administration of its data processing operation not only to enhance the systems we can build, but also to unify and enhance the system development process itself. New and powerful techniques for representing human knowledge, as well as advanced prototyping techniques, make this kind of merger possible over the next several years.

2.3 Methodology Overview

In this section and those that follow, we look in more detail at the activities depicted on Figure 1-3, Preferred AI Project Life Cycle, referencing the other life cycle diagrams for contrast.

The main line of technical work on the project begins with a Survey, or feasibility study, whose primary task is to decide whether or not a project is worth doing. When the project is approved it moves into Structured Analysis, in which we try to specify the user's needs in terms of functions to be performed and the data relationships among them. The Structured-Specification that results from analysis is used in the Design phase to specify how the user's needs are to be implemented, and the Implementation phase is devoted to constructing a production system as specified in the Packaged-Design.

During structured analysis, the logical information needs of the project—knowledge and data—are identified without particular concern for how they will be implemented in the physical world. This Logical-Knowledge-Base (KB) description is used in the Physical KB Design to specify the implementation details of the knowledge base.

The proper placement for Activity 9, Knowledge Acquisition, is somewhat enigmatic since during both the survey and structured analysis we actually collect quite a bit of expert knowledge. After analysis, however, design activities are likely to focus on software design issues rather than on collecting knowledge, and the knowledge acquisition process continues somewhat more independently. In addition, the maturation of the knowledge base to an expert state may extend well beyond program development.

Knowledge acquisition is an independent activity whose Knowledge-Base output is available to all other parts of the development process. In practice it is likely to be closely cooperating with the survey and analysis and becoming more independent after analysis. This suggests that it might be done by personnel on the survey and analysis team during those activities, and that those people might continue to focus on knowledge acquisition after analysis.

The Systems Integration and Hardware Analysis activities are certainly important to the life cycle although not always a direct part of the technical specification cycle. We shall deal with each of these subjects subsequently as they are affected by the unique characteristics of AI.

2.4 So What's New?

There are some major differences between the traditional structured life cycle and the AI project life cycle although they are not dramatically evident on the life cycle diagram. The differences are more in the way each phase is conducted, the people and processors which may be needed to do the AI component, and the content of the specification. For example, Knowledge Base Design, Activity 3 on Figure 1-3, deals with design of the physical knowledge base using a logical knowledge base specification from structured analysis. In the traditional structured life cycle of Figure 1-1, Activity 3 deals only with database design.

Such differences are amplified as we look at some details of the life cycle in the following sections.

3. Survey (Activity 1)

With clients who have little or no previous exposure to AI, I strongly recommend taking some time up front to learn about the strengths and weaknesses of the AI tool kit and formulate a short-term strategy—which frequently involves beginning a single small AI project—thus laying the groundwork for a long-term policy for integrating AI tools.

3.1 Goals of the Survey

Feasibility studies historically have taken very little time, a matter of a few days or weeks at the most; their main purpose has been to decide whether or not a specific project is worth doing and to provide some general goals and cost estimates for the project. This may continue to be true for knowledge-based systems once you have a specific project in mind. At the beginning of a company's exposure to AI, however, it may be worthwhile spending more time to understand the technology and to review many possible applications before choosing one.

If you already have in mind a knowledge-based project, then the survey may simply amount to applying the guidelines developed in Chapter 2, Selecting Expert System Applications. If you really have very little idea of where to begin, then the following discussion of the survey is directed at you.

When done in depth, the survey of Activity 1 addresses corporate strategy and application selection. Its method is to study a company's business and data processing environment for the purpose of proposing short- and long-term strategies for the smooth integration of AI technology. The survey results are intended to be a primary input to the company's internal process of deciding how and where to use AI.

By means of the survey we expect to determine, at relatively little expense, which domains are suitable for the application of AI; as I mentioned, Chapter 2 examines application selection issues in considerable detail. If there appear to be suitable domains, a second objective is to prepare a detailed proposal for carrying out a development project at least through the analysis phase.

The recommendations of this proposal are based on in-person interviews with executives, other management, and experts in various domains. Through the interviews, plan to get a good idea of the relationship of AI to departmental and corporate goals. Doing so makes it possible to identify one or more domains in which AI could provide substantial benefits in terms of increased revenues, reduced costs, solutions to presently unsolvable problems, or improvements in situations whose present solutions are less than optimally effective.

3.2 Major Result of the Survey

The expected result of the in-depth survey is a report recommending steps to be taken by a company over the short and long term in integrating AI into its business and data processing operations. The recommendations propose specific areas which could benefit from the application of AI as well as potential evolution in software, hardware, and training policies.

Within each potential application domain, the report may include the following:

- A narrative overview of the relevance of knowledge-based systems and, in some cases, high-level Data Flow Diagrams showing knowledge-based systems with their interfaces to the rest of the company's environment. (Data Flow Diagrams (DFDs) are the primary graphic tool used for system specification in a structured analysis. DFDs are described in more detail later in this chapter and in subsequent chapters.)
- A payoff/cost analysis of the application of AI to the domain.
- A risk analysis of the proposed development.

3.3 Additional Results

Some domains may appear to be particularly good candidates for knowledge-based systems. For those applications which exhibit the highest potential, the payoff/cost and risk analyses are done in more detail[3] and we may undertake some preliminary prototyping of the application.

3.3.1 Define Scope and Output

The survey is our first attempt at saying how a knowledge-based system will look when it's finished and how it will be used in production. It means considering specifically what the system inputs and outputs are to be, i.e. its data context and what decisions are to be made by the system, or what specific planning the system is to do. There are sure to be many refinements before the system is finished, but this first cut provides enough information to estimate the ensuing structured analysis activity.

It's likely that the information gathered here will be recorded using the same tools we shall use in Activity 2, Structured Analysis: Data Flow Diagrams with accompanying structured English descriptions and a project dictionary. By starting system specification as information becomes available during the survey, we have a good start on the structured analysis activity.

3.3.2 Begin Knowledge Base Definition

The knowledge base is the medium through which a human expert's knowledge is made available to the computer. Evolving the knowledge base to an expert state begins in the survey and continues throughout the project and beyond. The hope is that the knowledge base will grow eventually to outperform even the best experts in the subject domain. Though it may require many years to approach that lofty goal, within a reasonable time the system can become a useful, cost-effective apprentice.

The process of building a knowledge base begins and continues by watching experts make decisions and distilling out of that performance the factors that go into making the decision. It is certain that interviewing experts is important, but in an interview environment people make decisions differently than they make them under pressure.

During the survey we expect to accomplish the very important job of identifying experts and deriving a rough set of decision factors relevant to their domain of expertise. We may also create some general rules as a start on the knowledge base; I call this a pre-prototype version of the knowledge base. In time, we may throw away this early decision structure and start again. Nevertheless, this preliminary work is necessary to begin to clarify the expert decision or planning issues involved in this domain.

[3] These subjects are treated further in Chapter 2, Selecting Expert System Applications.

4. Human Resources

Before proceeding with further technical considerations, it may be worth addressing the question of what kind of people we need to do a knowledge-based system project. This is one of the major apparent differences between DP and AI: the human resources needed to do the job. Here is another case where I have translated AI jargon into concepts that are more common to traditional system development environments and, I hope, made them more readily apprehended by business and DP people.

During both the survey and analysis stages, the principal system development players are analysts, now sometimes called "knowledge engineers," and users, now called "domain experts." Why do we need these new terms to refer to functions that have been done for a long time?

4.1 The Analyst

Analysts have always tried to figure out what knowledge certain users have in a particular area—that's what system analysis is all about—and generally they seek out users who are expert users in that area. Still, AI people probably spent a long time thinking up the term knowledge engineer. Since they're certainly not just trying to make their special skills seem mysterious, there must be some significant difference between analysts and knowledge engineers, as there must be between users and domain experts.

The distinction, however, is a result of differences in quantity rather than quality. For example, what we expect from a domain expert is typically a much more detailed dump of that person's expertise than we usually get from "ordinary" users. One reason we expect more is that we have knowledge engineers who know how to do more with what the domain expert knows. But here again it's more a matter of having new analytical techniques than it is of knowledge engineers being qualitatively different from analysts.

At least let's hope so. If it is *not* the case that any competent analyst/designer can learn, fairly easily, to do knowledge engineering, then AI's future in business and industry is likely to be very short and uninteresting. At this early stage in the transfer of AI technology from lab to lunch counter, very few people know the techniques of knowledge engineering, and most of those are Ph.D. researchers in fields related to AI. To complete the technology transfer, essential theory and techniques must be distilled out of the base of research knowledge and organized in such a way that it can be used readily by analysts of quite different experience and educational backgrounds.

There are, of course, serious questions about where to start looking for people suitable to do knowledge-based systems work. My answer to these questions is that you start at home, in your own company. Many companies already have an untapped resource of people with a home-grown understanding of knowledge-based systems. Mainly, although certainly not exclusively, these people are based in the manufacturing, engineering, or research departments of a company and have been exploring AI systems on their own for many years. Since until recently this sideline interest has not been acceptable or useful in their work environment, they may have kept it in the closet.

4.2 The User

We should also say a few words about domain experts, although they are discussed in more detail in Chapter 2. People appear to gain expertise over a long period of time by building a large number of rules of thumb—heuristics—for dealing with new situations in their area of expertise. The kind of person we need as a domain expert is one who is clearly better at some task than others who are not considered to be expert.

For domain experts, knowledge base creation is an intense exercise in self-awareness. During this process we think, plan, formulate general hypotheses, test, refine, and start again more than once. To expect less than this is unrealistic.

The experts assigned to a knowledge-based project can continue their daily work, but they should expect frequent interruptions from the analyst. These cautions notwithstanding, knowledge base development is often a rewarding experience for the experts.

5. Structured Analysis (Activity 2)

Once an application has been selected, it is time to define completely the user requirements of the finished system. The stated purpose of an expert system project is to convert a human expert system into a computer expert system. To specify this conversion, the techniques of structured system analysis are very effective.

The primary outputs of an AI structured analysis are a structured specification of the system and a prototype knowledge base. It's difficult, without knowing the specific application, to estimate how much time the analysis phase will take. For an application in which almost all the analysis work is with the knowledge base and the expert task involves a single well-defined, thoroughly understood decision, analysis could take as little as three months. On the other hand, if multiple decisions are required, or if the decision/planning process is not well-known, or if a substantial amount of additional AI or DP functionality must be specified, then analysis could easily take six months or more.

5.1 The Structured Specification

The structured specification is a document which represents the final stage of an analytical process in which we look at the complex human expert system as the human implements it, and partition that large task into smaller mini-tasks, or mini-systems, which are easier to specify and implement. In fact one of the main goals of structured analysis is to study a big problem and partition it into mini-problems which are easier to understand. Each mini-system is then described with a written mini-specification. The structured analysis activity involves the following steps.

5.1.1 The Current System

A traditional DP system project usually involves adding new capabilities to an existing manual or automated system. To do this we need complete, unambiguous, and non-redundant documentation of what's being done now, so that changes can be integrated easily. With expert systems it is certainly possible to add new capabilities, but the main focus is on capturing what the expert does now and simply automating that process. Documenting the current system is a two-step process.

Step 1 is to document the expert's work exactly as it is being done presently. This means that if the expert uses a dozen file cabinets and a host of encyclopedic texts, then we document the use of those tomes as it relates to the expert's work. This is called documenting the **current physical system.**

Step 2 is to distill from the current physical documentation both the explicit decision or planning functions the expert is doing and the specific information (*not* documents) used to perform those functions. When finished this represents what we shall call the **current functional system.**[4] It represents the expert's work devoid of all considerations of current processing technology; we assume a technologically perfect world in which any functions can be done in any way without regard to efficiency or similar operating details. This is an important step, since our real goal in developing an expert system is to preserve the functional essence of what the human expert does and embed it in some new technological environment.

The current physical and current functional documents go hand in hand, and these steps tend to overlap considerably. The real goal is the current functional representation, so frequently the current physical documentation will be short-cut to get to the functional.

To exemplify the difference between physical and functional data and processes, consider the expert handling of expense claims. A salesperson returns from a trip and writes actual expenses on a paper form provided by a human secretary. The form is then given to a human supervisor, who signs the form and sends it to the accounting department. Accounting writes a physical reimbursement check and sends it to the salesperson. This is a physical description using physical data—expense forms and checks—which are handled by physical processors—people. It doesn't sound terribly expert!

A more functional description would show the expense information, not the paper form, going to an Approve-Expenses process which does require some expertise. The approved expense information, not the signed form, would go to a process called Prepare-Reimbursement which might output something called Expense-Payment. Though we don't yet know what the expertise is, this level of description allows the possibility of some expert decision work in approving expenses and preparing reimbursements.

It is, of course, the functional description of the expert's work that we are interested in when developing a knowledge base; the specification of those expert functions is the

[4]The jargon of structured analysis uses the term *functional* interchangeably with the term *logical.* I shall try to stick with functional to avoid confusion with the symbolic logic underpinnings of AI.

specification of the knowledge base. Documenting the current physical operation can give some important help in discovering the structure of that expert knowledge.

5.1.2 The New System

When we are reasonably sure of the current functional system, we can move on to specifying the new system. This is also a two-stage process whose goal is to get to a specification of the new system, including the technology to be used in running it.

Step 1 is to add any new functionality or data which may be needed in the new system. In this step, we are still working at a functional level, without consideration of how the expert system will be implemented technologically. There may be no new functions to be added to the current functional system—in such a case, the specification of the existing functions simply becomes more detailed, and we can move on quickly to the new physical system.

The point of persisting at a purely functional level as long as possible is to be as clear as possible about *what* we wish to do before considering *how* we're going to do it. This stage of the analysis documentation is called the **new functional system.**[5]

Step 2 is to specify the physical details of the new system and create documentation of the **new physical system.** With the complete specification of expert functions developed in Step 1, we are now in a position to re-introduce considerations of technology. This means deciding whether the expert system will operate interactively or in batch, whether it will involve centralized or distributed processing, and other decisions about how the system will work in the real, technologically imperfect world. We now have a structured specification of the knowledge-based system.

At every step in the structured analysis, documentation consists of a set of DFDs, with narrative descriptions of the processing depicted thereon, and a project data dictionary, an informational description of *all* data required for the system. The structured specification describes the final system completely from a technical point of view and is the primary input to the design activity.

5.2 Support Tools

During analysis, the structured documentation traditionally is developed on paper, and the result of analysis is a paper model of the system. More recently, however, there has been pressure to develop a working prototype of the system as early as possible in the project life cycle. There are now computer tools[6] being developed which allow not only the drawing of DFDs or DFD-like diagrams on the computer, but also may have some capability to prototype the DFD functions using generalized business processes built into

[5] For an example, refer to the DFDs in Chapter 5 (Figures 5-1a through 5-1d), which are roughly at the new functional stage.

[6] Examples of such systems are "Excelerator" by Index Technology of Cambridge, MA, and "The Analyst's Toolkit" by Yourdon, Inc., New York. There are also more sophisticated knowledge engineering environments such as IntelliCorp's (Menlo Park, CA) "KEE" and Inference Corp's (Los Angeles) "ART" environments.

the development tool.[7]

This ability to create approximations to a solution—prototypes—is particularly important in an expert system project both because knowledge base creation is a highly interactive process, and because the only proof we have for correctness of the knowledge base is in its ability to make valid expert decisions. There are now many domain-independent **expert system shells**—that is, generalized inference engines—which are excellent for at least beginning the construction of complex knowledge bases. Such systems have the ability to reason, make inferences intelligently, but have no knowledge about which to reason. They may or may not be used in the production version of the system, but they can be very useful beginning in the survey and continuing through the design phase.

During analysis, we expect to come close to a final version of the decision structure and function of each mini-system in the AI component, and hope to evolve those mini-knowledge bases to a point where the prototype system consistently handles at least the simplest real-world cases successfully. The latter goal will be accomplished more or less depending on the extent of the analysis. The completion of analysis also includes setting standards of acceptable performance for the system when it goes into production.

6. Design (Activity 4) and Implementation (Activity 6)

Although discussing the details of the design and implementation activities is outside the scope of this book, it is important to put such issues in context with analysis. Here is an overview of some of the current issues in AI system design and implementation, and a discussion of some alternative solutions.

6.1 The Goal of Analysis

The analysis activity bears roughly the same relationship to design and implementation as science does to engineering. The analytical specification represents somebody's good idea about what can be or needs to be done, and may even include a working prototype. Design and implementation take that good idea and make it work in the real world.

A long-standing goal in system development is what might be called **automatic programming,** or auto-magic programming as some of my colleagues call it. The idea is that we write a system specification in some very flexible descriptive language and then have the operating system run that specification directly, or at least automatically translate it directly into production program code. In short, the goal is to automate design and implementation to the greatest extent possible, thereby reducing the project's reliance on human activities.

One current attempt at automatic programming is the use of ad hoc query systems, particularly natural English query systems. In such an environment, each question is actually a very small system development project in which the English question is a

[7] The use of techniques for creating early prototype systems is discussed further in Chapter 6.

flexible, high-level specification of what we want to have done. Such a "system development project" includes no design or implementation activity; these are included as part of the underlying system, and simply entering the question causes the system to figure out how to do it.

At present the kinds of systems that can be developed using an ad hoc query facility are relatively simple and constrained. One can imagine, however, that over the long term we might be able to do all our system work in this same way. From this point of view, a goal for structured analysis would be to turn over the DFDs and their mini-system specifications, written in some high-level descriptive language such as English, to the operating system without further development. Unfortunately this is not to be for some time although, as we shall see in Chapter 6, there are now ways to directly implement prototypes of DFDs.

6.2 The Current Role of Design

During analysis, the work we do is oriented toward specifying the logical content of what it is that the user or domain expert wants to have done, *not* toward any particular algorithmic way of doing it. Further, the analysis documents describe the system as though it were a network of independent processors running simultaneously—that is to say in parallel—and controlled by the data relationships between them. Typically, these documents describe fairly accurately the actual situation in a human work environment. In the arena of computer systems, however, we do not presently have in widespread use operating systems and programming languages that understand this kind of **data-driven** network organization whose processes are specified by logic statements.

The purpose of the design activity traditionally is to transform the structured specification in two ways. First, it must produce documentation which specifies not only what program modules must be written, but also the algorithms to be used for those modules. Second, it must specify the architecture of the program modules in such a way that current operating systems, such as VM, CICS, or PC-DOS, can execute the system; typically this architecture is a hierarchical organization of program subroutines.

6.3 Design and Implementation Problems

This role for the design activity has worked relatively well for the kinds of commercial applications we have done traditionally. However, we are now entering a kind of never-never land where design is concerned. It is a time when the old hierarchical, procedural programming languages, like COBOL and PL/I, and the traditional hierarchical operating systems are not really as well-suited to knowledge-based applications as are the network DFDs of structured analysis. It is regrettably also a time when logic programming languages, such as PROLOG, and parallel processing hardware and operating systems are not even off the ground in terms of commercial availability.

6.4 Current Solutions

For now, there seem to be no fully satisfactory solutions to this quandary, although help is on the way from many quarters. There are alternative intermediate solutions, and the one to use in any particular case depends on many considerations, including the extent of financial commitment a company is willing to make to knowledge-based systems. Here are some of the alternative ways of dealing with this situation.

6.4.1 Ignore the Problem

Design and implementation are activities which presume that some computer software is being developed. For a knowledge-based system this software is typically related to building the part of the system that actually does the logical reasoning and planning—the **inference engine**—and any interfaces to more traditional software, to the user, or to databases.

Since the expert system shells being marketed today are becoming commercially credible from a software engineering point of view, one alternative is to use a shell without modification. It is very important, however, to be sure that its capabilities truly fit your specification in terms of not only interfaces, but also inference capabilities. Otherwise you may be unwittingly building constraints into your project that can prohibit the system from ever becoming truly expert in the domain you have chosen.

In addition, a class of **knowledge-based products** which have sophisticated natural language communication capabilities, inference components, and an extensive built-in knowledge base is beginning to appear.

This solution has the major advantage of avoiding the problems of design and implementation altogether. The price one pays, of course, is having little or no options in tailoring the software to a particular environment.

6.4.2 Minimize Change

If an expert system shell or knowledge-based product is not adequate by itself, then we must indeed worry about design and implementation.

From a business point of view, where maintainability of systems is a prime requirement, a viable alternative may be to simply continue the structured design[8] and implementation processes in their present form. It is certainly possible to design knowledge-based systems in a traditionally hierarchical way and to implement them in PL/I, FORTRAN, BASIC,[9] or even COBOL.

The advantage of this approach is that no new languages or hardware need be

[8] Among the excellent texts on the subject of structured design are: Page-Jones, M., *The Practical Guide to Structured Systems Design,* Yourdon Press, 1980; Yourdon, E. & Constantine, L.L., *Structured Design,* Yourdon Press, 1978.

[9] Some commercial expert system shells are written in these languages, as is the natural English query system, INTELLECT, which is written in PL/I.

obtained, and the training of personnel is entirely in the conceptual techniques of AI rather than in new programming languages. Where necessary during development, a commercial expert system shell can be used satisfactorily to prototype the knowledge base. In fact, some commercial systems[10] can be integrated readily with more traditional systems for use in some production environments.

In short, the DP and AI components are developed together, using the same resources with minimum new cost, and in a familiar environment.

Although this is a very conservative approach that could wear out its welcome in a fairly short time, a further advantage is that when you decide to move on to more advanced hardware and software technology, you have people trained in the conceptual technology already.

6.4.3 Maximize Technology

Many computer hardware manufacturers are now marketing a new class of computer called an **AI workstation** or a **LISP machine.** This category of computer is unique in that its internal logic and memory structures are designed to efficiently execute programs written in the programming language LISP. Although LISP itself is a rather high-level language, for these machines LISP is equivalent to assembly language on more traditional computers. LISP machines were developed in response to LISP's status for 30 years as the de facto standard programming language for AI research work.

An AI workstation, however, really has little to do with hardware. In fact, traditional mini and mainframe computers as well as some of the more robust microcomputers can be used as AI workstations, although somewhat less efficiently than a LISP-based machine. The AI workstation concept has more to do with the system development software being written for them. In particular, a variety of programmer-friendly **knowledge engineering environments (KEEs),** whose entire mission is to simplify the development of knowledge-based systems, have been written.

Thus another design and implementation alternative, which impacts analysis as well, is to grab technology at its leading edge and base the development of knowledge-based systems on an AI workstation environment. This alternative has considerable merit, but also involves substantial costs and risks.

The costs are related to buying new hardware and software, which can be over $150,000 for a single user workstation; hiring qualified AI technicians, whose starting salaries may be in the range of $60,000 to $80,000; and training current employees, whose productivity drops to zero during training and whose training costs may exceed $300 per day. The prices of all AI products and services are dropping rapidly as vendors' initial investments are paid back and the technology becomes popular. Nonetheless, at this point a substantial investment of money and time is required to take full advantage of the potential of AI.

Similarly, the risks involved can be prohibitive. For example, AI vendors are working

[10] Notably the KES system from Software A&E & TIMM from General Research Corp. of McLean, Va.

hard to develop efficient high-speed interfaces to more traditional system environments which will allow their hardware and software to appear to an IBM mainframe as simply another peripheral device like a disc drive. However, a truly practical capability is not yet available, and it's not clear that this can be done effectively. This means that if the company's human resources and payroll system with large databases runs in a traditional IBM environment, we may not be able to integrate a knowledge-based tax system that was developed and runs on a LISP machine.

All of these cautions notwithstanding, new hardware and software are beginning to make an impact in commercial companies. It's unlikely that a LISP machine will ever replace the IBM mainframe for running the company payroll, but **knowledge-based peripherals** may very well take hold and come into widespread use in business.

6.4.4 Buying AI

For some companies, buying AI capabilities may be the most appropriate path to follow, a path which involves using traditional DP techniques for the DP component and buying more advanced system development for the AI component. This approach is being used by many companies now and is reflected in Figure 1-2, Current AI Life Cycle.

For a time in this scenario, the AI and DP components probably go separate ways. In fact, different design phases may be proposed for each of them. The DP component is subjected to a structured program design and, since typically there is no in-house capability for AI programming, the design activity includes seeking quotations for any AI software work which needs to be done. Without knowing the specific domain, it is difficult to provide even a rough estimate of the design and construction phases.

If a suitable general-purpose inference tool is chosen at the outset of analysis, it is possible that very little vendor contract work will be required to bring the system to production status. Nonetheless, the company's production environment is likely to require modifications, at least to the input/output facilities of the system. The changes are probably best done by the vendor, and the design activity will include obtaining quotations from the vendor whose product has been used during analysis or possibly from other vendors who may, by this time, appear to have a more suitable product.

This current uncommitted approach seems to me to be the most wasteful of all. It is true that your company will spend comparatively little on software and hardware to reap some benefits of AI. However, if somebody else is doing all the interesting work, then you learn very little and have no basis on which to build future knowledge-based system capabilities.

6.4.5 What Shall We Do?

Of course, there are an infinite number of ways to combine these options. Unless you have a truly exceptional background as a manager, you probably need some help in finding the right blend for your company. My recommendations are as follows.

Start small, but *start.* This is a time when every medium-to-large company should be doing something about AI. At the very least, one fairly high-level person in the company should track and evaluate AI technology, sponsor periodic presentations by vendors and experts in the field, and assiduously search out and evaluate possible applications. If you're not doing something about AI, your competitors probably are.

That is a much stronger recommendation than I usually make, but I feel strongly that too many companies are now drifting with regard to AI, waiting in the wings for somebody to tell them the "right" thing to do, and that such companies may eventually pay a heavy price for their wait time. Nobody knows the right thing to do, and if commercial companies don't start figuring out how this technology applies to them, the products and applications will continue to be directed by those who have very little experience with business problems.

AI applications currently are being directed by the attitudes and experience of AI researchers whose business is AI research, not bottom-line profitability. I believe that companies such as yours have a responsibility to themselves, to their stockholders, and to the business community at large to figure out where AI is useful and where it is a waste of time. We are at a crossroads in AI where commercial companies are the only entities who can realistically determine the future of AI and the direction of applied AI research.

I must add that I have no fundamental quarrel with AI researchers. Their work over the last 30 years has been magnificent, but in a business environment they are fish out of water. The typical AI researcher has never worked in a company where the accounting books absolutely had to be closed by the third day of the following month, or where customers go someplace else when they can't get an answer to a product support question within a few minutes, or where people are fired because they can't get systems done on time, on spec, and on budget.

There is now a small but growing class of independent consultants, not necessarily large consulting firms, who have extensive experience working as employees in business situations as well as the technical and experiential background to understand AI. You may be able to find such a resource in-house, and that is all the better.

Having worked for several years as an independent consultant, I am well aware of many companies' bias against using such outside contractors. However, if ever there was a right time to draw on such people, this is it. The kind of person you are looking for understands not only traditional system development, but also the day-to-day concerns of business managers and the current marketplace for AI products.

Your goals in using such outside sources must be well understood. They are probably not going to do a knowledge-based system for you; they probably won't even make decisions for you regarding the right applications. What they will do is help you realistically assess your goals and decisions in the light of an understanding of this new technology. They can also help you find the right people, internal or external, to work on your knowledge-based projects, and they can evaluate the appropriateness of your specifications and designs in an AI context.

Following are some specific recommendations of how to get started.

Of first importance is attitude: let your beginner's mind have the freedom of knowing nothing about knowledge-based system development. In this way you can't help but learn from every new experience. Approach AI as you would any new project, looking carefully at the potential risks and benefits, and setting up milestones and checkpoints along the way, but try not to pre-judge experiences as they happen. Allow what you learn to guide revisions of your goals.

Second, consider your first project to be a throw-away. Whether you decide to build your own inference engine or use an expert system shell exclusively, encountering the subtleties of knowledge-based system development on a small scale is likely to save you a lot of grief in the long run. In addition, in spite of my urgings to so something soon, you do have time to be careful. The "AI industry," if there is such a thing, is in its infancy and you are far better off getting a firm grounding in the right techniques and procedures for your company through a demo system than you will be by starting off with a multi-million dollar project that has a relatively small chance of success.

In summary, look at this new technology the way you would assess a new business opportunity. Be prepared to do more than one round of financing, develop explicit intermediate milestones, and look for the big payoffs 3-5 years down the road based on careful planning, development, and marketing.

7. Knowledge Acquisition (Activity 9)

Knowledge acquisition means getting knowledge of a particular domain from some source, usually human, and building it into a computer system. This activity is where most of the effort goes in building a knowledge-based system. The question to be addressed is: How do we do it? The answers to that question are unequivocally unique to each application, and they have been addressed as a part of virtually every other subject in this book. Nonetheless, some general thoughts about knowledge acquisition may help.

In a certain sense one cannot separate knowledge acquisition from the rest of structured analysis. It is the interactive part of structured analysis in which analyst and user work together to build the logical, business-oriented structure of the DFDs and their mini-specs. One might indeed say that knowledge acquisition and structured analysis are one and the same thing. The major difference is one of convention: we shall call knowledge acquisition that activity in which we specify the functional, logical content of the expert's domain. Structured analysis, by contrast, will include the functional components of peripheral activities, such as user interface, and also specification of the physical or technological components of the system.

Let me emphasize that this is a completely artificial separation. Knowledge acquisition is in fact a part of analysis, the part which is the domain essence of the current functional and new functional system descriptions.

Knowledge-based system technology has separated, physically and logically, domain knowledge from the software that uses this knowledge for reasoning. One of the main benefits of this separation is the freedom it gives us to continue the knowledge acquisition part of analysis long past the end of the software system development, and to do so relatively independently of software development.

The Preferred AI Project Life Cycle shown in Figure 1-3 acknowledges this dichotomy in analysis, but makes it evident in an important way. Traditionally, we have always done knowledge acquisition. It has typically been such a straightforward, well-determined activity, however, that we never thought of treating it as separate. Notice, however, that the structured methodology has always treasured the importance and independent integrity of the *functional* DFDs even though they have always eventually been completely embedded in the subsequent new physical DFDs.

Thus all we have done is shown analysis as the two pieces it really is. We might well rename the activities on the life cycle diagram to: (2) Structured System Analysis and (9) Structured Knowledge Analysis. Subsequently in this book, I have dealt with this split in different ways in order to indicate the kind of freedom one has in analyzing a knowledge-based system. In the main, however, structured analysis refers to software development and knowledge acquisition to knowledge base development.

8. Systems Integration (Activity 5)

Since the original assimilation of computers into business, no systems have had as much potential to change job descriptions as do AI systems. For this reason it makes sense to address system integration issues simultaneously with the beginning of analysis.

The specific issues of job change and training are dealt with in more detail in the next chapter.

9. Knowledge Base Administration

The knowledge base, which is likely to be made up of several mini-knowledge bases, is qualitatively different from a database, and its long-term maintenance is different from database maintenance. Nonetheless, the knowledge base is a store of information which will change over time as the business changes, and that process of change needs to be managed. My recommendation is to establish a corporate Knowledge Base Administration (KBA) function, at least as well-placed as Data Base Administration (DBA), responsible at least for monitoring performance of the knowledge base and controlling modifications to it. In many systems the knowledge base actually includes the database as well; in this case a single administrative function could be used for both.

9.1 Knowledge Base Performance

Performance monitoring is needed for at least two reasons. One is that we need to know how we are doing in training the system to make expert decisions. The other reason is that, although after a fairly lengthy maturation period the knowledge base tends not to change a great deal, knowledge in the expert domain in the real world frequently does continue to change, and sometimes rather dramatically.

Let's say, for example, that we have built an expert system with the knowledge of a high-performance stockbroker. It is difficult enough to capture that person's knowledge at any point in time with a given set of factors for deciding how to invest. The problem becomes even more difficult when we realize that once we have found a technique that works, it's likely that others will begin to use it, and before long so many others are using it that it doesn't work anymore. At that point, either we must modify the knowledge base by hand, or the expert system must be able to change its strategy and try something else.

Many real-world knowledge base environments are not as dynamic as the expert stockbroker, but still the environments do change and so must the system's knowledge base. In any case, a knowledge base needs to have its own performance standards against which it is measured at regular intervals, and it must be scheduled for maintenance when necessary.

9.2 Modification Control

Modification control needs to be managed in the same way that any production system is managed. That is, there are likely to be production, test, and maintenance versions of the knowledge base. Once a production version has been established at the end of system development, it is changed only when the replacement test version has passed consistency and performance tests tailored to the requirements of its domain. There may be many maintenance versions and perhaps even multiple test versions, but there is only one production version at a time.

10. Conclusion

Clearly, the development of AI systems follows a life cycle very similar to more traditional DP systems: we must decide whether or not a project is worth doing (survey), analyze the user's requirements (structured analysis), develop an architectural blueprint for the system (design), and then either build the system in-house or buy components which will work in a particular environment.

The differences between traditional DP systems and AI systems are often a matter of quantity rather than quality although there are qualitative differences. The major differences are in the knowledge base, which contains not only data, but also relationships between data; in the inference engine, which typically uses techniques not found in traditional systems; and in the people needed to do the job: the knowledge engineer and the domain expert.

In the following chapters we look more deeply into many issues surrounding the AI project life cycle as we discuss expert system application selection issues in the next chapter, and then follow a particular case study through part of its life cycle.

Chapter 2
Selecting Expert System Applications

From our overview of knowledge-based system development methodology in the last chapter, it should be clear that there are good and not-so-good applications of this technology. In this chapter, we take a closer look at some parts of the AI project life cycle by studying the selection of suitable applications of expert systems. Specifically, this chapter develops a set of guidelines for selecting good expert system applications.

The current enthusiasm for expert system technology is good in the sense that it is encouraging a giant step in our understanding and use of automation. Unfortunately, it is also focusing attention mainly on AI technology rather than on solving business problems. Looking for applications of expert system technology is the wrong approach. The right attitude is to identify problems that need solutions and then, if it makes sense, to apply expert system technology in solving those problems.

This approach will eventually surface once again as the novelty of AI wears off and companies become comfortable using the knowledge-based techniques as tools, rather than as ends in themselves. In fact, virtually any traditional business application can probably be improved by the selective introduction of heuristic processing. For now, however, the prevailing attitude is to find suitable problems for the AI solutions we have in hand. This is partly a defensible attitude since until new techniques are understood for themselves we cannot know where or how to apply them.

Whichever approach you choose, more and more people are looking for guidelines to evaluate the fit between any given problem and the techniques of AI. The success or failure of any knowledge-based system project hinges on selecting a good application area, carefully defining its scope and output, and choosing appropriate knowledge representation techniques and inference software to do the job.

For this study, we will consider three major dimensions of an application's suitability for heuristic treatment: worth, risk, and corporate culture. Several other factors which affect the choice of an expert system domain, such as teachability and administrative controls, will be dealt with separately.

The guidelines in this chapter are merely suggestions which will in most cases lead to potentially good expert systems. They are not guarantees of success, nor should they be followed religiously. There are exceptions to every guideline, and I try to point them out along the way.

It should come as no surprise that the factors we consider in selecting a knowledge-based application are basically the same as those we would evaluate in choosing any system project. In most cases, there are some important differences in selecting knowledge-based applications, and it is those differences I wish to emphasize.

1. Worth

The reason for doing any software project has to do with a payoff of some kind. In commercial ventures, the payoff is frequently measured in terms of profitability, the bottom line. The bottom line can be measured in various other ways such as revenue production, cost reduction, productivity, employee satisfaction, or others. With knowledge-based systems we are concerned with the same bottom lines, but the type of task and the magnitude of both payoff and cost tend to be different from traditional system projects.

1.1 High Payoff Tasks

In general *we are looking for projects in which a high payoff will result from automating professional-level decision and planning tasks.* The payoff is likely to result from increases in the speed or reliability with which those tasks are performed, or from the ability to perform tasks correctly which, though definable, are too large or too complex to be done consistently well by humans.

DEC's XCON system is an example of the latter kind of improvement. XCON was the first, and for several years the only, expert system in daily production use. Its job is to assist salespeople at DEC by providing working configurations in response to customer orders for DEC computers. Before XCON was built, it was unusual for a humanly configured VAX computer to perform correctly after installation. Other applications—such as communication network diagnosis, financial planning, or investment management—may become more manageable when the expert's knowledge has been codified and can be applied consistently.

Carefully selected knowledge-based applications should have, and are likely to have, a higher payoff than traditional DP applications. One reason for this is that knowledge-based applications tend to address those problems which have not been dealt with previously because they were too difficult to handle with traditional methods. As a result, it may have been necessary to hire several highly skilled, and expensive, professionals to do the job.

1.2 Expert System Cost

Another reason to prefer high payoff applications is that knowledge-based systems tend to be very expensive to evolve to an expert level of performance. In general the reason for the high cost of knowledge-based systems is not the inference software which, conceptually at least, is fairly straightforward. The major expense is in acquiring the experts' knowledge to build the knowledge base.

For several years to come, the construction of a knowledge base is likely to continue to be a brute-force joint effort between domain experts and one or more analysts. By brute force I mean that there are no systems which can effectively interview a domain expert to capture knowledge, and no systems able to learn effectively from their own experience. Brute force means iteratively applying a trial-and-error approach to training the system to behave like the expert.

To get an idea of how expensive this is, consider the following situation. An average-sized application—say insurance policy underwriting—will take at least one year to achieve junior expert status and probably two to three years to mature. During that time, the services of at least one analyst are required full time. Knowledge engineer-type analysts may cost from $15,000 to $20,000 per month, and a $50,000-per-year expert's productivity can easily be reduced by at least 50% during knowledge base construction. Thus, the effective cost of a knowledge base which requires three years to mature could easily exceed $500,000.

The cost of such systems has already decreased dramatically since the early research systems were done and will continue to fall. These reductions are coming from several quarters: improved knowledge engineering tools, experience in building knowledge-based systems, and less-expensive knowledge engineers, to name a few.

1.3 Expert Application Generators

There is an exploding class of software system whose job is to assist analysts and experts in the process of building knowledge bases. I prefer to call such systems expert system shells, or application generators or even, more in keeping with the jargon, knowledge engineering tools. At the time of this writing, these tools are at best primitive—they generally sport a fairly user-friendly front-end and give their users the capability of teaching the system through general rules or specific training cases. They help by displaying the structure of the expertise or the rules being accumulated, will check logical consistency of the knowledge base, and can explain how decisions were reached. A very few of them[1] go so far as to suggest where their training is weak and try to generalize from the knowledge they possess.

None of the offerings so far gives the user any help in deciding what the decision or planning structure should be, nor are they able to learn on their own. These are serious deficiencies which, when alleviated, should significantly reduce the effort required to build expert systems.

In spite of their present shortcomings, these tools can greatly reduce the cost of an expert system project. They are generally fairly well debugged, are sold with training in their use, and their capabilities and limitations are well documented. Building one's own knowledge-based tools is bound to involve a lengthy analysis, design, and programming job, at least on early projects. In addition, severe penalties in training are exacted, as discussed in Section 2.3 of this chapter.

1.4 Other Worth

This focus on dollar-related payoffs notwithstanding, there are other reasons for doing an expert system that are related to worth. They range from applications which are crucial to keeping the business in business to those which may be done for demonstration, practice, or image only. Very broadly I like to categorize these reasons as follows:

- Crucial to the viability of the business.
- Necessary to solve a particular problem.
- Useful and profitable, but not necessary.
- Suitable for demonstration.

For some time to come, we can expect a rather high, though diminishing, cost for doing knowledge-based systems work. Thus one guideline is that *we seek applications which have a high payoff/cost ratio, or can guarantee some other result which is important to the business.*

1.5 Partial Solutions

As an addendum to evaluating the worth of an AI application, keep in mind that expert systems are not necessarily an all-or-nothing proposition. That is, not being able to deal with all of the expertise in a particular application may not rule out that application.

For example, we do not necessarily want to rule out stockbroker expert systems, or the art of selling expert systems, just because it may be too difficult to capture all possible expertise in those domains. We might be able to work effectively in those areas by automating only those pieces which have the characteristics approaching our ideal, while leaving the rest to human judgment. Over time, as our ability to represent and process knowledge improves, the pieces left out of the expert system at the start can be added.

In any case, the extent to which the problem can be solved by an expert system is

[1] TIMM, for example, a product of General Research Corp.

going to affect its worth. If applying expert system technology to a stockbroker's knowledge will solve only 10% of the problem, then clearly such a solution will be worth very little to us.

2. Risk

It is hardly necessary to provide business people with a guideline such as *choose applications that present low, or at least manageable, risk;* this comes naturally to most experienced business people. The more interesting subject is where the novel risks lie in knowledge-based applications.

The fact is that, regardless of what specific risks we choose to discuss, much of the risk arises ultimately from the fact that at this point there is very little commercial experience with knowledge-based systems. It is, in fact, difficult to find commercial vendors who have successfully installed even one production knowledge-based system, let alone any who have a successful track record. Everyone is still a pioneer in a certain sense; nearly every serious near-term application will involve substantial risk. The problem then is to look at the specific areas where trouble may arise and to provide guidelines for managing the risk.

Risk means the likelihood of not being able to complete a project to a point where the results, in terms of payoff, justify the cost to date. Risk is more the result of several other factors than it is a factor in itself. For example, if domain experts will not be available for the duration of the project, then the application is significantly at risk. The risk is that before they become unavailable, we may not be able to capture enough of their expertise to have a useful knowledge base.

Specific factors likely to affect risk in a knowledge-based system project include knowledge base complexity, inference technique complexity, and the ongoing availability of the domain experts.

2.1 Nature of Expertise

The list of potential applications for expert systems is very long: tax return analysis, insurance claim processing, risk management, problem diagnosis, and training, to mention just a few. The "goodness" of any of these areas depends partly on factors having to do with the nature of the knowledge involved, its reliability, completeness, ambiguity, and stability.

2.1.1 Scope of Knowledge

In the ideal situation, we would like to have an application in which knowledge is restricted to a very narrow domain[2] and decisions are entirely determined by measurable factors

[2]MYCIN, one of the first successful attempts to commit expert knowledge to the computer, restricts its expertise to the very specialized medical area of bacterial blood infections.

that don't change over time. That is, the easiest expert systems are those in which we know all the rules ahead of time and they are all deterministic.

The reasoning behind such an ideal is that the more firm our understanding of the expert domain is, the more reliable the system's decisions will be. On the other hand, a completely well-specified problem has little need of today's knowledge-based technology; we can solve such a problem with a traditional COBOL program.

In the real world, however, nearly every human expert makes decisions based on data which are to some extent incomplete, unreliable, ambiguous, and dynamic. It is for this kind of situation that expert systems are most applicable. The ideal of well-defined knowledge stated above is simply a goal toward which we strive. The ideal does, however, provide a useful measure when comparing two potential expert system applications.

For example, let's consider an expert system that underwrites insurance policies and one that teaches the art of selling. It is almost intuitively clear that in terms of our ideal the underwriting system would be quite good, while a system on the art of selling would be, at best, difficult. That is, while the underwriting task involves a relatively well-defined "YES/NO" decision as its output, teaching the art of selling is going to be quite different for every trainee.

2.1.2 Expert Eclecticism

A related problem, which sometimes makes it difficult to maintain a narrow focus of expertise, is that experts tend to be eclectic in their choice of information to make a decision. A stockbroker, when deciding what issues to trade, may bring into that decision information that has nothing to do with the financial aspects of stock trading. If some brokers are into astrology, for example, they may decide that at the time of the full moon one should consider trading certain issues. The point is that in such a situation, the expert system must be aware not only of stock prices, earnings, and other technical indicators, but must also have some expertise in the area of astrology.

2.1.3 Intuition and Common Sense

A corollary of a narrow, well-defined domain is that the decision or planning task should involve as little intuition and common sense as possible. However well an inference engine may deal with uncertainty in its knowledge base, these systems are still fundamentally deductive and so they leave a great deal to be desired when it comes to reasoning in nondeductive ways.

In summary then, the guideline regarding the nature of the expertise is that *we seek applications in which the decision or planning task is restricted to a narrow area, is relatively well-defined, and the data on which it is based are relatively complete, reliable, unambiguous, and stable.*

2.2 Domain Experts

The guideline regarding domain experts in selecting applications sounds relatively straightforward: *There must be at least one source of expertise, usually a person, who is provably better at the task than non-experts, and who is willing and able to be active on the project long enough to develop the knowledge base to maturity.*

2.2.1 Number of Experts

Regarding the number of experts, some feel that the ideal is to have only one expert. In a certain way this is the ideal situation since multiple experts, given the same set of facts, may reach different conclusions and introduce another level of uncertainty into the knowledge base. On the other hand, many real-world decisions are performed by different people, many of whom are expert in the field. Not to consider the variances engendered by differing opinions is to ignore large parts of the expertise that is actually used. One thing is clear, however, and that is that there must be at least one source of expertise —otherwise there is no human expert system to automate.

We may also wish to consider that some decisions are made by groups of people either sequentially, or as a group. Thus the actual number of experts required may vary from situation to situation. In any case I would recommend starting with as few experts as possible—otherwise the amount of uncertainty about the decision process may become overwhelming when all you're trying to do is to set some widely accepted ground rules for the domain.

2.2.2 Proving Expertise

This raises the question of what constitutes expertise, and how one goes about proving expertise. This is relevant not only to selecting applications, but also to assessing the maturity of the knowledge base as it develops. It's difficult to provide specific guidelines for determining expertise outside a specific domain, and in very many cases it is intuitively obvious who the experts are in a given area.

The process of knowledge acquisition may begin even before we approach a potential expert in a domain to determine if that person is expert or merely very good at a task. It is foolish to proceed with an expert system project without some relatively concrete way of measuring performance. The specification of such a **performance metric** requires that we know at least what it means to be expert in a domain. That in itself is domain knowledge although it may not be expressed as heuristic rules.

Rightly, our original guideline urges that experts be *provably* better than others, so we should in every case try to find some objective measure of expertise. For the insurance underwriter it could be objectively related to the size of paid claims versus premiums collected, for example. For the expert stock broker we could probably work up a formula based on average portfolio growth, income, and others.

Even if you are going into an area which previously has been too difficult for humans—that is, where there are no experts—you must find some way of determining whether or not you have improved the system's expertise by adding any particular rule to its knowledge base.

Naturally, the metric devised for measuring expertise is also likely to be used in determining the application's payoff.

2.2.3 The Expert's Attitude

What do experts think about expert systems? This differs widely from project to project and from expert to expert. In any case, the expert's attitude is crucial to the success of a knowledge-based system project. The expert's attitude is conditioned by many things, mainly the perceived worth of the system to the expert.

Some experts are overjoyed at the prospect of no longer being the only one in the company who can do a particular task, or in knowing that their expertise will not be lost if they retire; the advent of an expert system is a freeing experience for this kind of person. Such an individual is a joy to work with because the adventure of building the knowledge base is a truly cooperative and fulfilling experience both for expert and analyst.

Diametrically opposed to this attitude are those experts who feel that not only their job security, but their very worth as a human being, is based on their unique ability to do a particular task better than anyone else. To take away that task, or even suggest that a computer might be able to do it, becomes a personal insult and is often met with overt hostility. If such an attitude cannot be changed unequivocally, such a person should not be engaged to develop the knowledge base. I emphasize the need for a clear change of attitude because it is frequently possible apparently to dispel such an expert's overt hostility by applying various organizational or social pressures. The unfortunate result of this approach is simply to push the hostility underground where it will do even more subversive harm than when it is being expressed openly.

There is of course a broad range of attitudes between these extremes and, when trying to evaluate domain experts for a potential application, I have found it useful to categorize them broadly as willing and able, uninterested, incompetent, or hostile. The expert being **willing and able** is, of course, the desirable situation. Anything less than that kind of participation by the expert will make the project both more lengthy and more difficult to complete successfully. The **hostile** expert is to be avoided if at all possible.

The **uninterested** category of expert is peopled by those who just don't believe the job can be done or, if it can be done, that the result will be very useful. They are not likely to be cooperative in the active willing and able sense, but they will probably not do anything to subvert the project; they are just too busy to bother. It is OK to work with this kind of person, at least at the beginning of knowledge base construction when it is fairly easy to identify relevant constructive rules. As the project moves into the more difficult nitty gritty which truly differentiates expert from non-expert behavior, these folks are likely to resist work sessions with the analyst.

Things are not quite as bleak as they may seem with this type of expert. Often as the project moves along, it becomes evident to the uninterested expert that the system can in fact be useful and they may well become something closer to the willing and able type.

An **incompetent** expert does not do the task badly—such a person simply has great difficulty with self-observation and in explaining how the task is done. This dramatic lack of ability for introspection may not become apparent until the project is well underway and can have a crippling effect on knowledge base construction. It places an unbalanced responsibility on the analyst to become expert in the knowledge domain and to work almost exclusively from observation of the way the expert does the job.

Naturally, attitudes are not as clear-cut as I present them, and other characteristics of experts certainly affect the project. In addition, attitudes may change over the course of a project; it is incumbent upon the analyst to be aware of these changes and adapt to them.

2.2.4 Availability of Experts

Even with the "right" number of provably expert, willing and able people to start the project, it is doomed to failure if those people are not available long enough for the knowledge base to mature. People, companies, and jobs change in the real world, often with dazzling speed, and before an application is begun we must be reasonably sure that our domain experts either will stay with the project or that they can be replaced with other experts; this is one cogent reason for choosing applications with multiple experts.

We do expert systems for many reasons. Among them are the innate shortcomings of people: we're emotional, we get tired, we're inconsistent, and we tend to leave the job either by retiring, quitting, or dying. Sometimes the imminent or feared departure of an expert may be the very reason for starting an expert system project. Employees of long standing who have developed a depth of expertise over their careers may be candidates for replacement by an expert system as they near retirement age.

The advice here is to start the project well before such an employee's retirement, so that it can be fairly complete before the employee leaves.

In addition, we must consider how much of the time the expert can be made available. As I suggested earlier, we acquire the expert's skill primarily by interacting with that person while the expert task is being done expertly. And so taking the expert off the job for a series of interviews is silly and counterproductive at best; the expert stays on the job.

However, the expert now has an encumbrance: the analyst, whose job is to be an expert nuisance. The analyst must in some way codify the expert's work and so is constantly interrupting for explanations, generalizations, alternatives, and so on. As a result, the expert's productivity can easily be reduced by 50%.

2.3 Build or Buy AI

Expert application generators offer some choice in making the buy-or-build decision, and the cost will vary depending on that choice. At this time, a company should probably begin with one of the commercial products if it embodies technology adequate to solve the problem. If it does not, then two other options are to seek modifications from the vendor, or embark on knowledge-based system development in-house.

Bringing knowledge-based system development in-house represents a long-term commitment to changes in training, software, and possibly computer hardware as well. For example, although knowledge-based techniques are often quite simple once they are learned, training traditional systems people to be maximally productive with the software techniques can take 2-3 years.

In any case, part of the risk involves whether or not adequate inference software is available to do the job. If not, it will have to be a modification of existing software or built from scratch.

3. Corporate Culture

Without trying to provide an anthropologically accurate definition,[3] let us say that "corporate culture" refers to those structures, values, and behaviors which are generally observed in and expected by a particular group, in this case a company. The guideline here is that *the expected results of building an expert system should be consistent with a particular corporate culture.* For example, in a company where policy is dictated by the president without apparent regard for its effects on individuals, an application which results in the wholesale elimination of jobs may be appropriate. However, in a company with more participative management, such an application might be regarded as suspect.

3.1 The New Age

We are well into an era in which the old style of dictatorial management is becoming less and less productive. This is partly because individual growth now ranks high among the priorities of people in the work force. A remarkable number of people would rather not take a job than work in an environment which is not conducive to their own growth as individuals. This attitude has been reinforced by the entrance into the executive suite of people from the '60s generation who wish to consolidate the shift in attitudes that was painfully begun in that decade.

[3] I recently co-led a workshop entitled "Computers and Corporate Structure" whose goals were to discuss the ways computer technology has affected organizations and to identify issues needing further research. In addition to representatives of academic sociology and anthropology, I invited executives from large and small companies as well as some therapists, consultants, and journalists. A major part of our discussion focused on whether what we in the business world call "Corporate Culture" is in fact a culture. It became clear that regardless of academic accuracy, the popular use of a term generally overrides the rigorous definitions made by specialists in related fields.

Every individual has a guiding vision of the future, a career, a mission in life, and every company has a similar vision which guides its actions toward some future. To the extent that these visions reinforce each other, it makes sense for a person to work in a particular job for a particular company. The new age company and the new age employee are aware of this situation and can work together cooperatively as long as it makes sense to do so. It is to the benefit of both of them to part company when their paths diverge significantly.

Naturally it takes considerable expertise, both from an individual and from management, to monitor this interplay reliably over a long period of time. One might even consider building an expert system to assist in this job, though at this point the complexity of such a knowledge base is prohibitive.

Realizing that employee satisfaction is a major component of profitability in the long term, many large and small companies are providing benefits which encourage an employee to explore self-fulfillment in the work place. Preventive health benefits, meditation rooms, and programs which focus on individual growth are among these new perqs.

To some it may appear that such benefits are merely another way of manipulating people for the sake of money, and for some companies this may be true. But there is a growing realization that supporting individual growth for its own sake is the long-term road to corporate success.

3.2 Can Expert Systems Replace People?

It has been suggested elsewhere in this book that expert systems possess the potential to replace people. This is actually nothing new—computer systems have been causing job descriptions to change for 30 years or more. What's different about expert systems is that we're now talking about replacing professionals, not clerks.

Some professional-level tasks requiring specialized intelligence can actually be done as well or better by an expert system as by human experts. In some cases this is a desirable feature, while in others it is a subject for real concern. An insurance underwriting expert system could replace a large percentage of a company's underwriters, and this may be desirable both for the underwriters and the company since those people could frequently be doing even more useful tasks elsewhere in the company.

For the time being, however, it is better to think of an expert system more as an apprentice, an intelligent assistant, than as a replacement for human experts. This approach is more palatable and more realistic in the near term. Certainly no expert system is going to replace anybody until its knowledge base matures, which can be a matter of several years. And in many cases, we would like to have humans present for their humanness as well as their expertise: physicians or receptionists are examples of this.

On the other hand, companies such as LMI[4] are deeply involved with expert systems for real-time process control in petroleum and other industries. Although LMI admits to

[4] LISP Machine Inc., a subsidiary of Texas Instruments, is one of the leaders in the manufacture of computers oriented toward AI applications. They share this market with several companies including Symbolics, Inc., Apollo Computers, and Xerox.

a long-term goal of replacing human operators, they acknowledge that in the foreseeable future, the only sensible approach is to provide an intelligent computerized assistant for the operator. Should they or others succeed in reaching their long-term goal, large numbers of people could be left without work in their main career skill.

3.3 A Reason for Caution

Through AI, we are embarking on a course which, if followed to its ultimate potential, can challenge every corner of our awareness of what it means to be human. Even as I sit before my word processor with all of my awareness of how infantile our present science of AI is, a very small part of me wonders whether some day (well beyond my present lifetime) a robot could sit in this same place and write these same words. For those who are less aware than I am of the limitations of current technology, this wondering can grow to become an insidious drag on enthusiasm and productivity in a job.

Expert system technology forces us to examine closely the relationship between computer capabilities, our feelings of uniqueness as humans, and our individual worth to a company's vision of its future. For this reason, the advisability of doing an expert system project must be considered in terms of its effects on the individual, likely changes in the corporate culture, and hence the long-term health of the company.

For purposes of this application evaluation process, we will lump all of these thoughts under this guideline: *Choose applications for which employee acceptance of the system is positive.*

4. Other Indications

There are numerous other indications of situations which are right or wrong for expert system treatment. Some of these are presented below and can be stated fairly briefly with little explanation.

4.1 Adequately Solved Problems

I frequently encourage people not to be satisfied with adequacy. Where expert systems are concerned, however, the magnitude of the task of building a mature knowledge base is sufficiently expensive and risky that I prefer to capitulate to the tried and true maxim: "If it works, don't fix it."

You can determine the expertise of the expert system by comparing it with the performance of a more traditional equivalent system. For demonstration purposes, therefore, a currently solved problem may be worth attacking. Nonetheless, unless there is a significant payoff available through the expert system approach, adequately solved problems can probably wait.

4.2 Teachable Expertise

Earlier in this chapter I mentioned the incompetent expert as being one who was not aware enough of the expertise to be able to explain it. This issue goes beyond self-awareness in that there are skills, such as the art of selling, which are very difficult to teach and seemingly require some innate ability if a person is to become expert. In a case where expertise is very difficult to teach, application of expert system technology is counter-indicated.

Conversely, if a skill is routinely taught to new employees such as underwriters, then its teachability suggests that it may be a good application.

4.3 Turnover

Situations in which there is a fairly high turnover of experts are good candidates for expert system technology. Plant operators are good examples of this, as are underwriters and technical diagnosticians.

4.4 Easier Solutions

While AI has been advancing for the last 30 years, so has data processing, and in many areas DP has provided satisfactory solutions to complex problems which could be addressed also by an expert system. In Chapter 4 we look at several potential applications. The one we finally choose is an expert seminar scheduling system which might be more easily solved using the mathematical optimization techniques of linear programming. In this case the best solution involves a marriage of the old and new; in some other cases, it may be better to stick with techniques which are known from the past to be effective.

5. Conclusion

We now have a fairly long, and perhaps confusing, list of criteria for deciding whether or not a particular application is suitable for a heuristic approach. Furthermore, these are only guidelines, so we may not be 100% sure once we have made a decision about an application. Perhaps we need an expert system to help decide whether or not a potential application is a good one. It is, in fact, considerations of complexity and indefiniteness such as these that often suggest that a knowledge-based system might be appropriate.

Whether or not you decide to build an expert system to help select expert system applications, selecting good applications of expert systems is crucial not only to the success of a particular project, but to the long-term view of knowledge-based system work which develops in your company. I encourage you to take as long as necessary to find a *good* first application, probably small though visible, and one that is *useful* to your company.

The next chapter illustrates some of the principles of expert system development through an expert system called the Knowledge Base Evaluator (KBE), which can help us select appropriate applications.

Chapter 3
An Expert System Example

Starting with our list of guidelines from the previous chapter, this chapter looks at the Knowledge Base Evaluator (KBE), an expert system to evaluate expert system applications.

Although it would probably be too small an application for other than demonstration purposes, let's try to apply some knowledge base thinking to sorting out this application selection problem. I call this system the Knowledge Base Evaluator—KBE for short—and have given it substance through an expert system shell called TIMM.[1] In addition to providing some practice at building a knowledge base, we also have a chance to look at one of the knowledge engineering tools in action.

The process of knowledge base development takes place in three steps: specifying the decision to be made, finding the right structure of the decision factors,[2] and training the system by giving it general rules or sample decisions on which it can base its own decisions.

[1] TIMM, The Intelligent Machine Model, is developed and marketed by General Research Corp. of Santa Barbara, Calif. It runs on DEC's VAX computers, on IBM mainframes, on the IBM PC, and is fairly easily adapted to most other hardware.
[2] In AI jargon, "factors" are also sometimes called "attributes."

1. The Decision

First we must realize that in this case the expert task to be done is a decision rather than a planning task, and a fairly straightforward decision at that. We want to know whether or not an application with certain characteristics is a good expert system application. Not only is it a simple decision, but it doesn't require interfacing to any external databases or output other than the decision itself. This situation is easily handled by most expert system shells on the market today.

It's not quite that simple, of course, since there is a broad spectrum of suitability from very good to very bad. For this reason we might be tempted to specify our decision on a scale from 0 to 10, but be careful. The factors I have talked about in this chapter are general guidelines, often not very well specified, and in a real world situation they may not be known with great accuracy. This suggests that a 0 to 10 decision may be too fine a breakdown given the fuzziness of the factors that go into making the decision. A better choice would be to limit ourselves to a three-way decision: POOR, FAIR, or GOOD.

2. Decision Structure

Identifying the decision structure means that we watch and talk to the experts making the decision and from those observations specify both the attributes that are considered when making the decision and the values that these attributes can assume. We will take as our source of expertise the discussion in Chapter 2, Sections 1 through 4.

2.1 The First Cut

The Data Flow Diagrams (DFDs) of structured analysis are a good way to study an application and specify its knowledge base structure and content. In this case I have started in a more typical, ad hoc way to specify the knowledge and will soon convince myself that such an approach is less than satisfactory.

We discussed the goodness of an application in terms of worth, risk, effect on corporate culture, and other indicators. These then could be the factors we're looking for—if we can assign some metric to each of these factors, the system should be able to provide us with a decision. Figure 3-1 shows one possible decision structure based on these assumptions.

This says that we are trying to decide on something called Suitability, and it can be either GOOD, FAIR, or POOR. Further it says that we need consider only four factors in reaching the proper decision, and each factor has only a few values. If this were the final structure of the knowledge base, then teaching the system to be expert would mean making up rules such as:

```
DECISION:             VALUES:
      SUITABILITY:      POOR     FAIR     GOOD

FACTORS:              VALUES:
      WORTH:            NEGATIVE       LOW           MODERATE   HIGH
      RISK:             LOW            MODERATE      HIGH
      CULTURE:          NEGATIVE       NEUTRAL       POSITIVE
      OTHER INDIC:      NEGATIVE       NEUTRAL       POSITIVE
```

Figure 3-1. First Cut KBE Decision Structure—SUITABILITY

```
IF WORTH IS HIGH
   RISK IS LOW
   CULTURE IS POSITIVE
   OTHER INDIC IS POSITIVE

THEN SUITABILITY IS GOOD
```

Using this structure, if we could specify each factor value definitely, then we could completely specify the expertise in this problem using 108 specific rules. That is, there are only 108 possible combinations[3] of values, so even for a demonstration this is too trivial to be useful. We could simply write down all the combinations on paper and look up the result as needed.

However, we do expert systems because of their ability to reason with uncertain data, and it's likely that in any real case we might not know the worth, risk, or effects on corporate culture with certainty. It's more likely, for example, that we would assess the worth of an application to be 50-50 MODERATE or HIGH. As soon as we admit uncertainty into the decision process, the complexity of any decision becomes much higher and we may well need an expert inference engine to deal with only four factors.

In addition, the inclusion of OTHER INDIC as a single factor seems awkward since "Other Indications" as discussed in Chapter 2 don't really hang together as a group.

2.2 Refining the Structure

Apart from uncertainty, the application is really more complex than four factors. We discussed several ways in which worth or risk could be measured, so that the value of these factors could be the result of a second level of decision. WORTH as shown in Figure 3-2, for example, is actually the result of a PAYOFF/COST factor, a PERCENT SOLUTION factor, and a TYPE factor. This means that when asking the system to make a decision

[3] In a decision structure such as this, the number of combinations of values is equal to the product of the number of values each factor can take on.

we would actually specify values for PAYOFF/COST, PERCENT SOLUTION, and TYPE, but not WORTH directly.

We could, of course, simply include these second-level factors in place of WORTH in Figure 3-1, but conceptualizing the problem is easier if we specify a two-level decision. One hypothesis which people use regularly and effectively is that a big problem will be more easily solved when broken into smaller, independent problems, assuming that we know the relationship between those smaller problems. Commercial expert shells should allow a general networking of decisions so that the value of any factor may be the end result of many other nested decisions.

Another reason for grouping related factors into separate decision structures is to help limit the system's training and the complexity of its decision process (as discussed subsequently in Section 3.3 of this chapter). The idea is that to optimize its own performance, expert application generators sometimes do a factor analysis to determine related factors and factors that will rule out the most choices quickly. By structuring the decisions as we are for KBE, we are effectively doing some of that analysis for the system, which should be at least as reliable as the system would do based on its training.

The values chosen for each factor and the decision itself are more or less arbitrary in this example; no clear natural values emerge. In other expert systems there may be specific values or ranges which the experts use regularly in their work. In each case here I have tried to constrain each factor and decision to three or four values that provide a rough, but useful, bracketing.

Similarly, RISK could be specified in terms of the second-level factors we discussed in Chapter 2, Section 2. Listing all the base criteria—amount of intuition/common sense, potential for including partial expertise, and others—makes a fairly complex decision structure. It might be better to use a second decision level for RISK as we did for WORTH. The three main areas of risk discussed were: complexity, expertise, and administrative controls on data entering the system and compliance with procedures. Figure 3-3 shows the decision structure for the second-level decision concerning RISK, each of whose factors are the result of a more detailed decision.

This refinement of the decision structure is an intimate part of any knowledge-based system and deserves thorough treatment. It should be clear that in larger expert systems,

```
DECISION:           VALUES:
     WORTH:              NEGATIVE    LOW       MODERATE  HIGH

FACTORS:            VALUES:
     PAYOFF/COST:        UNDER 1     1-1.5     1.5-3     OVER 3
     PERCENT SOLUTION:   UNDER 50%   50%-75%   75%-90%   OVER 90%
     TYPE:               DEMO        USEFUL    NECESSARY CRUCIAL
```

Figure 3-2. Second Cut KBE Decision Structure—WORTH

DECISION:	VALUES:		
RISK:	LOW	MODERATE	HIGH
FACTORS:	VALUES:		
COMPLEXITY:	LOW	MODERATE	HIGH
EXPERTISE:	UNAVAILABLE	OK	AVAILABLE
CONTROLS:	LOOSE	OK	TIGHT

Figure 3-3. Second Cut KBE Decision Structure—RISK

this process of determining the actual factors and values used by an expert can be a time-consuming process.

The tabular specification format shown above can be confusing without some graphic way of showing the relationship between decisions. I prefer to use data flow diagrams (DFDs) to represent the factors and decisions, together with a dictionary to keep the values of the factors and decisions.

2.3 DFD and Data Dictionary Defined

Figure 3-4 shows a DFD which represents KBE's "final" decision structure, and Figure 3-5 shows the dictionary which contains the definition—the values—of each item of data on the DFD. This structured analysis specification technique[4] is widely used for specifying traditional data processing systems, and as you can see it is equally applicable to knowledge-based systems. In fact, this is the only technique I've seen which can reduce what appears to be a gigantic knowledge base to more manageable independent knowledge bases.

The DFD is a very simple and powerful notation for partitioning any big problem into smaller problems which can be solved independently of each other. Each circle represents a smaller problem, or **mini-system,** which works independently of the other mini-systems except for the data they share. Data passage is shown on a DFD by the arrows, called **data flows,** which connect mini-systems to each other. The boxes which border the DFD are called **sources** of data or **destinations** for data. They represent information generators or users whose processing is beyond the scope of the system we are specifying, but which give or take specific data.

The **data dictionary** uses the following simple notation to define the information content of every data flow on the DFD:

= An equals sign means the item on the left of the sign consists of, or is defined to be, whatever is on the right.

[4] A much more complete description of the techniques of structured analysis is contained in my book, *The Practice of Structured Analysis,* Yourdon Press, New York, 1983.

+ A plus sign in the definition means the defined item contains both what is to the left of the sign and what is to its right.

[. . .] Square brackets in the definition mean that the defined item contains one and only one of the items contained in square brackets and separated by a semicolon (;).

{ . . . } Curly braces in the definition mean that the defined item contains from zero to an infinite number of occurrences of whatever is between the curly braces. It is possible to precede the opening brace with a number indicating the minimum number of occurrences if not zero, and to follow the closing brace (}) with a number indicating the maximum number of occurrences if not infinite.

(. . .) Information contained in parentheses in the definition is optional. This is a convenient shorthand way of showing either zero or one occurrence—e.g., (Person) is exactly the same as saying 0 {Person}1.

* . . . * Information between asterisks is a comment about the definition, but is not part of the definition itself.

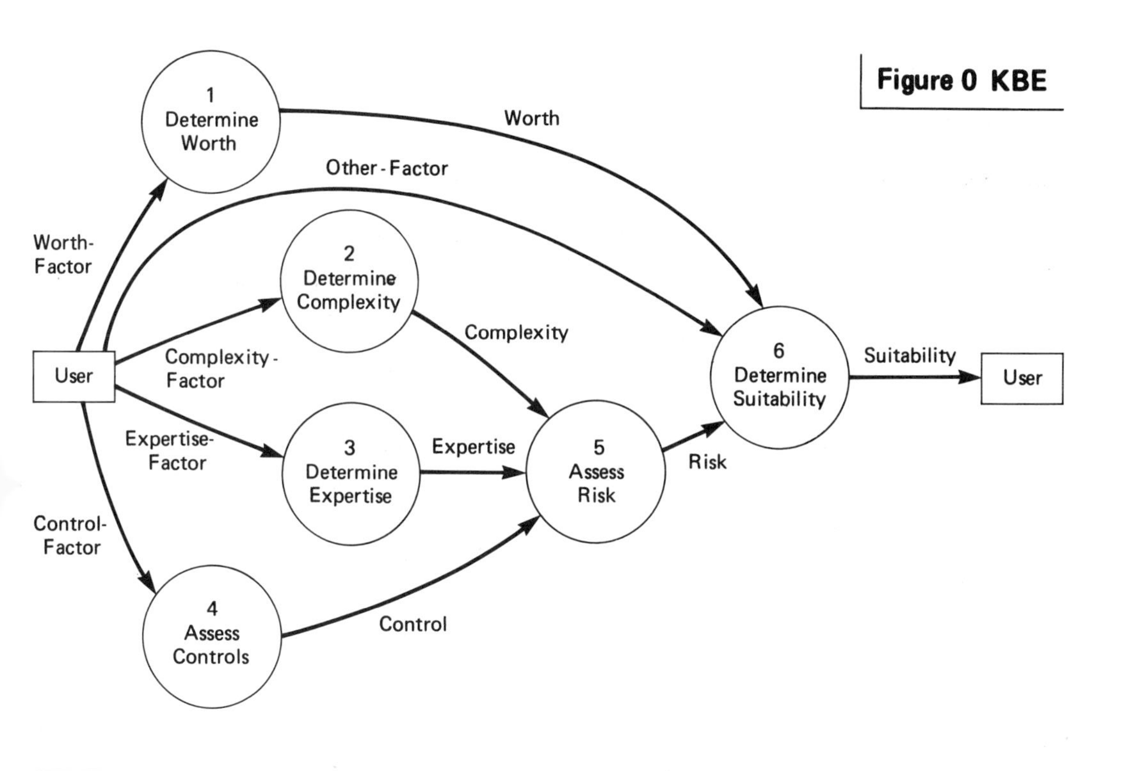

Figure 3-4. Data Flow Diagram for KBE Structure

Figure 3-5 contains several examples of the use of this notation. Items in the dictionary can be either data structures or data elements. Definitions in upper/lower case are further defined in the dictionary. Control-Factor, for example, is a **data structure** since it is defined in terms of three items which are further defined. Upper case definitions are values taken on by **data elements;** the values are not defined further. Data-Control, for example, is a data element of Control-Factor and is defined in terms of its values.

```
Complexity                = *Decision made based on Complexity-Factor.*
                            [LOW; MODERATE; HIGH]
Complexity-Factor         = *Factors related to application complexity.*
                            [Intuition/Common-Sense; Technology;
                            Decision-Definition; Knowledge-Domain]
Control                   = *Decision made based on Control-Factor.*
                            [LOOSE; OK; TIGHT]
Control-Factor            = *Factors related to administrative control.*
                            [Data-Control; Procedure-Control;
                            Performance-Metric]
Culture                   = *The effects of the system on corporate culture.*
                            *This is a synonym for Employee-Acceptance.*
Data-Control              = *Degree of control over data inputs.*
                            [LOW; MODERATE; HIGH]
Decision-Definition       = *How well defined is the decision?*
                            [FUZZY; OK; WELL DEFINED]
Easier-Solution           = *Is there an easier way to solve the problem?*
                            [NONE; PARTIAL; COMPLETE]
Employee-Acceptance       = *How will employees react to the system?*
                            *This is a synonym for Culture.*
                            [NEGATIVE; NEUTRAL; POSITIVE]
Expert-Attitude           = *What is the experts' attitude?*
                            [HOSTILE; INCOMPETENT; UNINTERESTED; WILLING]
Expert-Availability       = *For how much of the project will the*
                            *experts be available?*
                            [<50%; 50%-75%; 75%-90%; >90%]
Expertise                 = *Decision made based on Expertise-Factors.*
                            [UNAVAILABLE; OK; AVAILABLE]
Expertise-Factor          = *Factors related to the domain experts.*
                            [No.-of-Experts; Expert-Availability; Turnover;
                            Expert-Attitude]
Intuition/Common-Sense    = *How much non-deductive thinking?*
                            [<10%; 10%-50%; >50%]
Knowledge-Domain          = *How broad is the experts' knowledge?*
                            [ECLECTIC; OK; NARROW]
No.-of-Experts            = *How many domain experts are there?*
                            [0; 1-3; >3]
Other-Factor              = [Employee-Acceptance; Solution-Available;
                            Easier-Solution; Teachability]
Payoff/Cost               = *What is the estimated ratio of payoff to cost?*
                            [<1; 1-1.5; 1.5-3; >3]
Percent-Solution          = *How much of the problem will be solved*
                            *by doing this system?*
                            [<50%; 50%-75%; 75%-90%; >90%]
```

```
Performance-Metric    = *How good is the measure of expertise?*
                        [LOW; MODERATE; HIGH]
Procedure-Control     = *Degree of control over admin. procedures.*
                        [LOW; MODERATE; HIGH]
Risk                  = *Decision made based on Complexity, Expertise,*
                        *and Controls.*
                        [LOW; MODERATE; HIGH]
Risk-Factor           = [Complexity; Expertise; Control]
Solution-Available    = *How good is an existing solution?*
                        [ADEQUATE; PARTIAL; NONE]
Suitability           = *Final decision made based on Worth,*
                        *Risk, and Other-Factors.*
                        [POOR; FAIR; GOOD]
Teachability          = *How easily is the skill taught?*
                        [DIFFICULT; POSSIBLE; FREQUENT]
Technology            = *What are the technological requirements?*
                        [BUILD; ENHANCE; MODIFY; EXISTS]
Turnover              = *How high is the turnover of experts?*
                        [LOW; MODERATE; HIGH]
Type                  = *Why is the application being considered?*
                        [DEMO; USEFUL; NECESSARY; CRUCIAL]
Worth                 = *The decision made based on Worth-Factors.*
                        [NEGATIVE; LOW; MODERATE; HIGH]
Worth-Factor          = *What is the worth of the system?*
                        [Payoff/Cost; Type; Percent-Solution]
```

Figure 3-5. Data Dictionary for KBE Structure

2.4 A DFD Structure for KBE

Figure 3-4 shows the DFD for KBE. It shows that in spite of the apparent simplicity of this application, six independent decisions need to be made, as shown by the six mini-systems. Each of these receives its own unique factors and sends its decision on to another mini-system.

The DFD serves as a **model** of the decision process, which means that before programming anything we should be able to trace the result of any set of inputs to the system. Paper-checking the system in this way has been a valuable method in traditional systems development and is enhanced here if we are using one of the knowledge engineering tools. The use of such tools allows us to develop a prototype of the system very quickly and incrementally so that as the knowledge base develops it can be checked immediately.

For example, the DFD indicates that at some point the User, a source, enters something called a Worth-Factor which we see from the data dictionary is either a Payoff/Cost ratio, a Percent-Solution, or a Type. Using this information, the Determine-Worth mini-system will somehow, not yet specified, transform this input into an output item called Worth. We know, again from the dictionary, that Worth can be NEGATIVE, LOW,

MODERATE, or HIGH. We see also from the DFD that Determine-Suitability uses Worth along with two other data flows to output Suitability, the end result we are seeking.

All of the other user inputs can be traced, or walked, through the diagram to account for all possible inputs and outputs.

On this diagram I have grouped all of the factors that affect Worth into a single data flow called Worth-Factor, which is defined in the dictionary to consist of the data items Payoff/Cost, Percent-Solution, or Type. The [. . .] notation indicates that a user will input either Payoff/Cost, Percent-Solution, or Type, but not more than one of them at a time; the user may subsequently input the other factors. Similar groupings have been used for other inputs and help to control the complexity of the DFD itself.

This structure was input to TIMM so that it will know the factors and values to consider when we ask it to make a decision. TIMM's view of the KBE structure is shown in Appendix A. Our work in using TIMM is to input values for each factor related to the application. TIMM will then use its inference engine and its past training to give us the suitability of this application.

3. Training the System

However we decide to represent the DFDs, we still have the often major task of teaching the system to make decisions expertly. In expert system parlance, giving such rules to the system is often called **training the system.**

We set out to specify the way decisions are made concerning the suitability of a knowledge-based system to a particular application. So far, in drawing the DFD and writing the data dictionary, we haven't actually specified anything—all we have done is to break the big system into mini-systems and state the data interface between those mini-systems. We now have to write a mini-specification for each mini-system.

3.1 Structured English

In traditional structured analysis, mini-specs are written using a technique called structured English[5] and we will use structured English to specify KBE's mini-systems. A major difference is that while traditionally mini-specs completely specify the process to be done, in knowledge-based systems they may provide only a skeleton of the decision rules and some training examples. Filling in the gaps is left to the inference engine, which will apply the mini-specs in practice. That is, we are not telling the system completely how to make the decision, but giving it some guidelines regarding the experts' policy for making the decision. Some mini-specs will be more complete than others.

As an example, let's consider the mini-spec for the mini-system Determine-Worth.

[5] A structured English specification uses ordinary English, but restricts the description to equations or declarative sentences, objects which are described in the data dictionary, and the logical constructs of structured programming. See Keller, op. cit., for a more thorough discussion of structured English.

The goal of the mini-spec is to describe the transformation of the mini-system's inputs into its outputs. In this case there are only three factors coming in with four possible values each. This means that there are 64 possible combinations of values if they are well-defined. Uncertainty in assigning values may yield many more combinations, and the system's decision may have some uncertainty associated with it.

3.2 Training Guidelines

Assuming we can't or don't want to specify the decision completely, we need guidelines for deciding what rules or training cases should be given to the system, that is, written in the mini-spec. In general, the guideline is to *give the system rules that will help it come to a final decision quickly,* that is, by having to consider the least number of combinations. This could be done by ruling out many combinations with a single rule such as:

```
IF EXPERTISE IS UNAVAILABLE
THEN SUITABILITY IS POOR
```

It also helps the system to *specify combinations of values which are impossible* such as:

```
HAVING UNAVAILABLE EXPERTISE AND LOW COMPLEXITY
IS IMPOSSIBLE
```

Another help is to *give the system rules that lead to a definite decision.* There will be plenty of uncertainty in real-world examples—the more definite the system's training, the more reliable will be its performance in areas of uncertainty. Such rules often represent the limiting, either ideal or definitely bad, cases. For Determine-Worth we might well enter the ideal case where:

```
IF PAYOFF/COST IS >3
AND TYPE IS AT LEAST NECESSARY
THEN WORTH IS HIGH
```

A final guideline for now is to *give the system at least one rule for each value of the decision.* In Determine-Suitability, which can have the Suitability output values of POOR, FAIR, or GOOD, we should have at least one example of a POOR, FAIR, and GOOD application in the training.

3.3 How Much Training?

The total number of possible combinations of factor values is sometimes called the **decision** or **search space.** In KBE, if we tried to treat each of the factors in combination with the others, we would find that there are more than 1.6 billion[6] possible unique combinations of values, plus at least half again that number when we allow for uncertainty in specifying values in a particular case. Exhaustive training of the system is clearly out of the question. Our goal in training is to have the system consider as few as possible of those 1.6 billion combinations in reaching a decision.

In KBE, a dozen or so rules developed by applying the training guidelines given above can significantly reduce the problem very quickly. For example, saying "If expertise is unavailable, then suitability is poor" immediately reduces the combinations by a factor of 3. Systems such as TIMM also help themselves in this regard by being clever enough to consider first those rules which will most quickly reduce the job to a manageable size.

As shown by Figure 3-4, we have further simplified both the system's decision process and the rule development process by reducing the overall evaluation problem to six independent decisions—the DFD's six mini-systems. Now the number of combinations of values is the sum of the combinations of values for each mini-system. In KBE this means that the 1.6 billion has been reduced to a mere 1,600, thus reducing the complexity of the knowledge base by a factor of 100,000. From this example, you can see that the DFD partitioning process is not just to make the problem easier to conceptualize; the structure revealed by this partitioning frequently reduces the problem from one of astronomical complexity to one which is quite manageable.

Within any of those mini-systems, even for this small KBE example, the rule development process can still be complicated. Determine-Suitability, for example, uses up 972 of the 1,618 possible combinations, and so we still could use some help in developing rules. TIMM gives some of that help in the rule development process in another way. In terms of well-defined combinations of values, the decision space has a certain maximum size, as we have seen. By applying a fairly simple metric, TIMM has a concept of the logical distance from any one training case to all other combinations of values. Naturally, the reliability of the system's decision goes down as the distance between a new case and the closest training case increases. The concept of logical distance allows the system to tell us where its training is weak, and it does this by suggesting training cases for us to add to its experience. This relieves us of a great burden of complexity, considering that we tend to start getting confused when the number of things we have to keep in mind is more than about 7.

Appendix B lists six basic sets of rules developed with TIMM's help and used in evaluating several potential applications in the next chapter. Each of the sets of rules represents the mini-spec, or knowledge base, for one of the mini-systems on our KBE Data Flow Diagram. With this minimal training, TIMM makes relatively good decisions, but, as in any real-world example, there is always room for improvement. The guideline regarding how much training to give the system is to *start small and grow the knowledge base incrementally over time.* A very small knowledge base gives us only a start, but people are much better at taking something imperfect and making it better than doing something right the first time.

[6] 1,632,586,752 combinations to be precise.

4. What We Have Done

However simply, we have gone through the process of developing a knowledge base for what could be a useful application. It is certainly not complete in the sense of considering all factors which could go into selecting expert system applications, and by this time you may well have thought of things which could usefully be added—the likely number of rules, for example, as an additional Complexity-Factor. Many such additions are possible, but be careful. Just as it is possible to specify a decision too fine for the quality of knowledge, it is possible to try to capture the knowledge in too great detail. In this area, experience is the best teacher.

We do have a working tool, however, and shall apply it subsequently.

5. Including Inference

There appears to be one major omission from our KBE diagrams, and that is any reference to the part of the system which knows how to reason: the inference engine.

5.1 System Analysis vs. Knowledge Acquisition

In order to keep things in perspective, we must remember the split personality of analysis that was described in Chapter 1. That is, analysis is used on the one hand to specify inference technique and other system functions, and on the other hand to specify the acquisition of the expert's knowledge of a domain. Both types of analysis use the DFD notation, although for somewhat different purposes.

So far in this KBE example I have completely ignored the inference mechanism and have used the DFD notation exclusively to describe the structure of our knowledge base. This means that the mini-specs for the DFD of Figure 3-4 are likely to specify general rules or specific cases of decisions related to selecting applications, not how that knowledge is to be used by the inference engine.

5.2 The Role of the Shell

How can we get away with specifying the knowledge structure only, as in Figure 3-4, and not the inference software which will use it?[7] In fact, we cannot omit the inference component. However, an expert system shell, TIMM in this case, provides a universal, or global, inference engine for every mini-system in our example. For this reason, and

[7] I am indebted to Thomas Murphy of Taylor Instruments, a Division of Combustion Engineering, Inc. in Rochester, N.Y., for his insights that have helped to clarify my thoughts on appropriate structures for an expert mini-system.

because for this system we are not concerned with the workings of TIMM itself, we may decide to omit it from the DFD in the same way that we would omit the operating system from the DFDs that specify a more traditional application.

In other situations, however, it may well be important to see the role of the global inference software, and we might change the DFD of Figure 3-4 to that of Figure 3-6. In doing this we recognize that many expert system shells, as well as home-grown inference engines, are **interpreters**—that is, they can access program statements in a source language, such as BASIC or the IF-THEN rules in our knowledge base, and effectively execute them in that form. Although this is a much less efficient way to execute code than to compile it into machine instructions, interpreted code allows more freedom in changing the code as it is being executed. This kind of freedom is particularly handy in most knowledge-based systems.

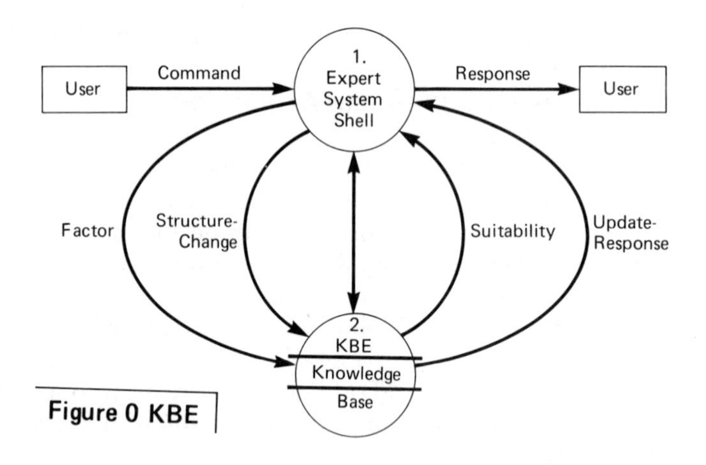

Figure 3-6. Physical DFD for KBE with Shell

As we can see from the DFD of Figure 3-6, the expert system shell is not only a global inference engine, but also is the piece that communicates with the end-user and accesses the knowledge base rules as though they are data: retrieving, executing, and possibly updating the logic as needed to respond to particular user inputs. The inference engine part of the shell has the job of deciding the order in which to invoke the rules based on the results of executing those rules.

5.3 DFD Knowledge Base Notation

Figure 3-6 introduces a piece of DFD notation which does not exist in traditional DFDs. It is a symbol for a knowledge base: the mini-system circle overlaid with the parallel lines that represent a data store. The purpose of this notation is to indicate the intimate mixture of knowledge as data with knowledge as process. Thus the knowledge base may have mini-specs which represent the logic of its process nature as well as data structure documents which represent its data nature.

When used on a DFD, it is typically acceptable to show only the information flow implied by the logic of the knowledge base, although it is more complete and correct to show the data access paths as is done here. The use of the knowledge base symbol implies what happens conceptually, and what may actually happen in a physical system: the needed knowledge base logic is accessed as data and temporarily becomes part of the inference processor code. The mini-spec for the inference engine includes such data accesses in addition to its other logic.

This knowledge base entity is important to see in some different situations:

1. In the physical DFD implementation of a knowledge-based system.

2. On those functional DFDs where the knowledge itself may be modified as part of the logic of the inference process.

3. In situations where some pieces of the knowledge base may have their own local inference requirements.

5.4 Local Inference Processors

If we were not using an expert system shell or some other system which can provide global inference and knowledge base management capability, then we may need to show the details of the inference processing and knowledge base management in different places on the DFD. In fact the possibility for local inference processors suggests that every knowledge mini-system may have yet another level of detail such as Figure 3-7, a closer look at Determine-Worth. This shows not only the Determine-Worth Knowledge Base, but in general may contain a local command interpreter, inference engine, and knowledge base manager as separate entities. Further, although it's not shown on Figure 3-7, the inference engine may have a control component which can heuristically modify the inference method itself.

There are also some functional DFDs such as Figure 3-4, in which the knowledge base may be considered simply as the functional logic part of the specification, so there is no need to indicate a separate inference engine or the role of knowledge as data.

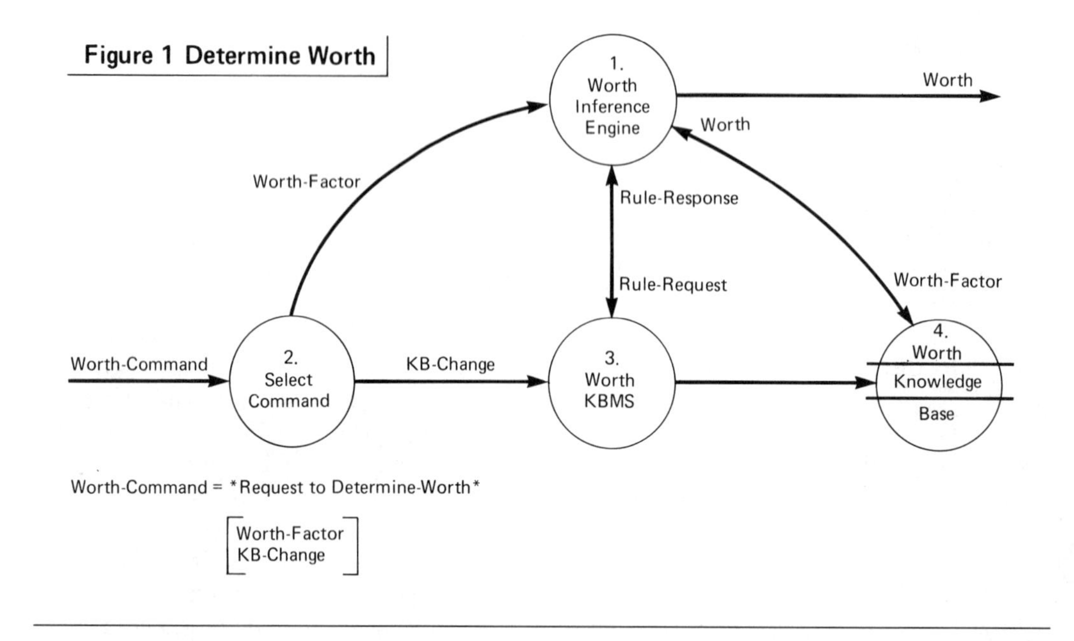

Figure 3-7. DFD Figure 1 - Determine-Worth

6. Conclusion

In this chapter we have used some of the guidelines for selecting expert system applications to build an expert system to apply the guidelines. This could in fact be an interesting and useful demonstration application in some companies, or it could expand to become a very complex knowledge base in some other companies. In any case, even if you don't have an expert system shell available, the structure of the decision, the dictionary, and the rules can easily be used manually in evaluating real world applications.

Selecting good applications for expert systems is crucial not only to the success of a particular project, but to the long term view of knowledge-based system work which develops in your company. Take as long as necessary to find a good first application, probably a small though visible one, and one that is useful to your company.

We are now ready to use the KBE to help us look at some other candidate applications.

Chapter 4 A Study in Application Selection

With the application evaluation guidelines provided in the previous two chapters, let's begin a study of a fictitious company, Generic Enterprises (Generic, for short), in its search to solve its problems and keep up with technology. This chapter looks at various potential applications of AI, much as we might do in the survey phase for a real company. Chapters 5 and 6 study structured analysis and prototyping issues for one application.

The main work of the survey is identifying and evaluating one or more potential applications. Although in an actual Survey this evaluation sometimes requires drawing high-level DFDs to get a handle on such things as complexity, I am going to put off further DFD work until the next chapter, which treats the issues of the structured analysis phase in some detail. Where necessary in this chapter, I shall make arbitrary assumptions to arrive at a suitability decision.

1. Generic Enterprises

Figures 4-1, 4-2, and 4-3 are memos and letters which provide some background on the situation at Generic. I should say in passing that most of the case material presented here is a blend of real-life experiences, which have been modified beyond recognition. If you recognize your company or problems being discussed in Generic's situation, this is purely coincidental.

Generic has problems in many areas of its diverse business, which includes seminar scheduling, communications network diagnosis, microelectronics sales, and insurance. This is certainly not a complete list of areas to which knowledge-based processing is applicable, but it covers a broad spectrum of likely candidates. Let's proceed by evaluating each area in terms of the criteria provided in Chapter 2, and then get the opinion of KBE, our Knowledge Base Evaluator expert system discussed in the Chapter 3.

GENERIC ENTERPRISES

To: Thurston Edwards
From: The Chief

SCATHING MEMO

We have a problem, Edwards, and it's in your department. Last week, two of those overpriced instructors of yours showed up in South Wales to teach the same seminar.

The week before that, I heard from the president of General Eclectic that NOBODY showed up to teach either of the seminars we had spent months selling to them. Furthermore, no one ever confirmed the dates and no one bothered to tell our "expert" scheduler that the scheduled instructor was on vacation for a month.

Also, last week you sent old Foobong to teach at Miscellaneous Management Services and *everybody* knows the people at MMS can't stand Foobong.

The list of foulups goes on and on! This is your last chance to exhibit a little intelligence, Edwards, and get your manual scheduling system up-to-date.

Figure 4-1. Generic Enterprises Memo 1

GENERIC ENTERPRISES

To: The Chief
From: Thurston Edwards
Mgr Education
(Now repentent and transformed)

I'm not sure why our scheduling system doesn't work any more. Maybe Jamie has too many people and events to deal with now? She used to do the job very well.

Having 35 instructors, each living in a different city, makes it tough to keep a running schedule of about 30 engagements per month. Last year I tried the linear programming system our management science folks use. It's OK for things like minimizing travel expense, but it can't deal with the fact that people like Foobong can only show his face in about three cities, or customers who won't accept anyone but Chauncey, or those seminar instructors who refuse to go to Winnipeg in January.

I've called the Renaissance man about our problem, and he said we might be able to use some high-tech stuff called Artificial Intelligence. I'm not quite sure what that is, but I remember your saying I need to use a little intelligence, so I'm going to try it out. By the way, he says it probably won't cost us more than about $1 million; what a bargain for all that technology!

He also suggested that if there were one application, there might be more and he would be glad to look over some possibilities with us.

Boss, I've turned over a new leaf. We're really moving into the 20th century.

Figure 4-2. Generic Enterprises Memo 2

GENERIC ENTERPRISES

Renaissance International Corp.
Two Deerfoot Trail
Harvard, MA 01451

Dear Renaissance,

My seminar manager suggested I write regarding some possible improvements we could make by using high-tech methods. Since you have spoken with Mr. Edwards, it should be clear to you that we have first-hand experience with *artificial* intelligence. In any case, here are descriptions of some other problem areas which seem to require extensive training and experience to achieve excellent performance. Perhaps you would be good enough to review these before you visit.

And by the way, I'd like to know why you've made each evaluation, and which others of our vast network of experts might benefit from your bag of tricks.

Communications Network

The single most expensive problem we have is diagnosing problems on our communications network; in fact, it costs us tens of thousands of dollars for every minute the network is down.

At first pass you might not think this is a big problem, since we have very sophisticated error-detection hardware at each of our 100 switching stations in cities across the country to monitor 30,000 miles of lines.

Also, only about 1% of the half million alarms we get each day from the switches and multiplexers are worth worrying about.

Still, we seem to have too much information and no way other than technicians' folklore for interpreting it. A single failure can cause legitimate alarms from several switches as well as sympathetic alarms from many others. There are a few long-experienced people who can do good interpretations quickly, but there are not nearly enough of them.

All we can do for now is to throw a mess of technicians at the problem and hope that one of them quickly discovers the real cause.

For the future we want to anticipate problems resulting from things like network deterioration so that we can re-route traffic ahead of time.

Negotiations

The worst problem in this area has to do with the microelectronics line we opened last year; it looks like we are going to have to close it down since the airheads in our sales department can't seem to negotiate a contract that doesn't wind up costing us money.

I suppose it's not all their fault, since in order to open this line I closed down another and brought the sales force over intact. They had done well in the past, selling packaged systems, but now they always wind up short when negotiating contracts with our customers; they never seem to know how to deal with the different personalities they meet, and they get outsmarted on subtleties.

At least we have old Hoskins who knows more about negotiating than any man alive. He does his best to give the younger guys some tips about what to look out for; he's worked with most of our customers' negotiators at some point in his 35-year career. Unfortunately, Hoskins will be retiring next year and all that savvy will be lost.

So, Renaissance, get your team to work and help us save some of these situations.

Confidently,

Ima B. G. Mogul

Figure 4-3. Generic Enterprises Memo 3

2. Seminar Scheduling

Generic is in the business of teaching seminars and needs to do some scheduling. As you can see, the scheduler is having trouble keeping up with the demand for changes in the schedule, although she is intellectually capable of doing the job in a professional, expert way. The fact that engagements, or instructors, or facilities may be added or cancelled on an hourly basis means there are just too many changes to be able to do the job well manually. Is this a good expert system application?

Figure 4-4 shows a list of the application evaluation criteria developed in the last chapter, and we shall use this format to decide what values to input to our TIMM implementation of KBE. The same form will be used for the other candidate applications subsequently. The list shown has underlined the values chosen for the scheduler project; I shall discuss only those choices that seem to raise important issues for a particular application.

```
Complexity-Factor = *Factors related to application complexity.*
      Decision-Definition      = *How well-defined is the decision?*
                                 [FUZZY; OK; WELL-DEFINED]
      Intuition/Common-Sense   = *How much non-deductive thinking?*
                                 [<10%; 10%-50%; >50%]
      Knowledge-Domain         = *How broad is the experts' knowledge?*
                                 [ECLECTIC; OK; NARROW]
      Technology               = *What are the technological requirements?*
                                 [BUILD; ENHANCE; MODIFY; EXISTS]

Control-Factor    = *Factors related to administrative control.*
      Data-Control             = *Degree of control over data inputs.*
                                 [LOW; MODERATE; HIGH]
      Performance-Metric       = *How good is the measure of expertise?*
                                 [LOW MODERATE HIGH]
      Procedure-Control        = *Degree of control over admin. procedures.*
                                 [LOW; MODERATE; HIGH]

Expertise-Factor  = *Factors related to the domain experts.*
      Expert-Attitude          = *What is the experts' attitude?*
                                 [HOSTILE;    INCOMPETENT;    UNINTERESTED;
                                 WILLING]
      Expert-Availability      = *For how much of the project will the*
                                 *experts be available?*
                                 [<50%; 50%-75%; 75%-90%; >90%]
      No.-of-Experts           = *How many domain experts are there?*
                                 [0; 1-3; >3]
      Turnover                 = *How high is the turnover of experts?*
                                 [LOW; MODERATE; HIGH]

Other-Factor      = *Factors not included elsewhere.*
      Employee-Acceptance      = *How will employees react to the system?*
```

```
                                    *This is a synonym for Culture.*
                                    [NEGATIVE; NEUTRAL; POSITIVE]
      Easier-Solution             = *Is there an easier way to solve the prob-
                                    lem?*
                                    [NONE; PARTIAL; COMPLETE]
      Solution-Available          = *How good is an existing solution?*
                                    [ADEQUATE; PARTIAL; NONE]
      Teachability                = *How easily is the skill taught?*
                                    [DIFFICULT; POSSIBLE; FREQUENT]

Worth-Factor        = *What is the value of the system?*
      Payoff/Cost                 = *What is the estimated ratio of payoff to
                                    cost?*
                                    [<1; 1-1.5; 1.5-3; >3]
      Percent-Solution            = *How much of the problem will be solved*
                                    *by doing this system?*
                                    [<50%; 50%-75%; 75%-90%; >90%]
      Type                        = *Why is the application being considered?*
                                    [DEMO; USEFUL; NECESSARY; CRUCIAL]
```

Figure 4-4. KBE Evaluation Criteria—Scheduler

2.1 Intuition/Common Sense

There appears to be a fair amount of common sense involved in deciding which of two schedules is actually better, and so it is likely that this system will be implemented as an apprentice to the scheduler rather than as a replacement. This suggests also that, because of the overburdening nature of the task at present, the human scheduler will look on the system as a blessed relief.

2.2 Easier Solution

Clearly we would like to automate the system, but consider the following questions:

Is scheduling really a professional-level task requiring intelligence?

If it is, then is building a heuristically driven expert system the best way to automate it?

The answer to the first question is Yes; there is little doubt that finding the optimal schedule of 35 instructors for a dynamic complement of 30 engagements per month is a monumental task requiring considerable intelligence. This takes the form both of a sophisticated awareness of what constitues the "best" schedule, as well as skill in manipulating instructors and engagements both on paper and in person.

The second question regarding the best way to automate scheduling is not so clear; after all, people have been using the techniques of Linear Programming (LP) to do scheduling for a long time, and apparently it works well.[1] So why not just throw some numbers into an existing LP model? If scheduling is an intelligent activity and we already have a general computational model, LP, which fits it, and if the goal of AI is to find general computational models of intelligent behavior, then haven't we already succeeded in applying AI to scheduling and many other activities requiring optimization?

Not so fast. There are some important restrictions on mathematical optimization problems, the most severe from our point of view being that the all the numbers in the function to be optimized must be known constants, and all of the constraints must be well-defined. In other words, the problem must be a well-defined, deterministic optimization. For many situations involving scheduling of production resources, these requirements can be closely approximated and so LP yields satisfactory solutions. In our situation, however, many of the criteria for assigning instructors or rescheduling conflicting engagements are not well-defined or deterministic, and so the methods of heuristic inference may be applicable after all.

I choose this scheduling example intentionally to illustrate the fact that it is more important to identify problems that need to be solved than to find places to apply heuristic techniques. Hardly ever is the application of knowledge-based processing an either/or situation; most systems can benefit from both knowledge-based and more traditional techniques. The effective approach is to find areas crying out for solution of some kind, perform a thorough structured analysis to identify fundamental pieces of the problem, and use whatever techniques will best solve each piece. It often happens, however, that in situations where a heuristic approach is partially appropriate, a knowledge-based orientation to the entire system will predominate, with more traditional approaches becoming servants to the heuristic parts.

In our scheduling problem, both heuristic techniques and search techniques such as might be used in traditional LP are applicable. It is this application which has been chosen for further study in the next chapter. At that time we shall examine the places and the ways in which heuristic and traditional techniques are interleaved.

We could make the following general guideline regarding what techniques to use:

IF you have a mathematical model, or other system,
that solves part of a problem well
THEN seriously (about .85 certainty)
consider using it where it fits,
and use a heuristic approach to solve the rest.

[1] Linear programming is one of many mathematical techniques for optimizing certain well-defined combinations of variables. It has been used often for dynamic production planning, as well as distribution and product mix problems. Mathematical programming is potentially a very powerful way of dealing with optimization problems. For further information, refer to Zionts, S., *Linear and Integer Programming,* Prentice-Hall, Englewood Cliffs, NJ, 1973.

2.3 Decision Definition

The output from this system is not a simple yes/no decision, but rather a matching of instructors to engagements, although each instructor-engagement pair is partly the result of simple decisions. The application actually involves a fair amount of planning to come up with an optimal schedule in which there are no conflicts.

Nonetheless, the criteria for deciding whether one schedule is better than another are fairly well-defined: a schedule is optimal when travel expenses are minimized and both customer and salesperson satisfaction are maximized. The fact that "satisfaction" is not numerically quantifiable does not mean that we can't provide guidelines for comparing situations.

2.4 Technology

There are no systems that I know of which take a knowledge-based approach to scheduling. There are, however, LP models which have been applied to similar scheduling problems. In this case I have chosen to say that we must build from scratch, since modifying the knowledge representation methods of an LP model are likely to be more complex than building some LP-like searching into a knowledge-based system.

This is often the case where knowledge-based techniques are blended with more traditional techniques. The new knowledge representation techniques are typically more flexible and powerful than traditional techniques and so, as I mentioned previously, the more traditional approaches tend to be subordinated to the new.

2.5 Data and Procedure Control

Control of administrative procedures is a bit flaky in Generic's seminar department, as evidenced by the scheduler not knowing when certain instructors are on vacation. Unfortunately, no amount of high technology will force an irresponsible contract instructor to let the scheduling department know about availability in a reliable way. This can be a serious problem in the real world and must be taken into account in evaluating this application. We are still very much aware of the garbage in, garbage out principle from DP antiquity.

We can legislate that instructors are to update their availability list weekly or daily and even provide simple mechanisms for doing this. But the best scheduling system in the world is seriously hampered if the procedures for controlling its external sources of information are not followed.

Control of data and administrative procedure control often go hand in hand, and in this case they do; inadequate control of procedures results in inadequate data.

2.6 Payoff/Cost and Type

In this situation, doing something about the problem appears to be crucial if the company is to stay in the education business. Often in such a case the Payoff/Cost ratio is difficult to assess and sometimes appears to be almost irrelevant. In training KBE to make decisions, we might well include a rule in Determine-Worth which says:

```
IF TYPE IS CRUCIAL
THEN WORTH IS HIGH (regardless of Payoff/Cost)
```

If there is such a rule, then the value chosen for Payoff/Cost is immaterial. If we don't know whether there is such a rule, then it's probably a good idea to tell KBE that the Payoff/Cost ratio is very high, >3 in this case.

2.7 Suitability

Figure 4-5 shows KBE's decision regarding the suitability of the scheduling application to expert system treatment. A desirable feature of the TIMM implementation of KBE is that whenever the system makes a decision, it gives us the choice of accepting or rejecting it. If we like the decision, it may also be a good idea to add that case to the knowledge base as a new piece of experience or training.

3. Using an Expert System Shell

Figure 4-5 is the first example we have seen of interaction with an expert system shell, and it's worth pausing to examine this dialogue. In general what's going on is that we have started the shell running with a knowledge base of rules called KBE, as shown in Appendix B, and what we see is its response to our having asked it to decide on the Suitability of a particular application.

3.1 Using the DFD

Typically, an expert system shell requires us to build a different knowledge base for each application. In this case, the KBE knowledge base consists of six even smaller knowledge bases, one for each of the mini-systems shown in Figure 3-4 of the previous chapter. In other words, the KBE expert system, whose final decision is the Suitability of an application, is actually made up of a network of six mini-expert systems whose interplay is controlled by the shell.

```
******************************************************************************
              © Copyright 1984 General Research Corporation
                TIMM (TM) The Intelligent Machine Model
                              Version 2.0
******************************************************************************

PAYOFF/COST                 IS over 3
PERCENT SOLUTION            IS over 90%
TYPE                        IS crucial

WORTH                       IS HIGH(100)
                               (Reliability = 100)

EMPLOYEE ACCEPTANCE         IS neutral
SOLUTION AVAILABLE          IS none
EASIER SOLUTION             IS partial
TEACHABILITY                IS possible

INTUITION/COMMON SENSE IS 10%-50%
TECHNOLOGY                  IS build
DECISION DEFINITION         IS well-defined
KNOWLEDGE DOMAIN            IS narrow

COMPLEXITY                  IS MODERATE(33)
                               LOW(33)
                               HIGH(33)
                               (Reliability = 41)

NO. OF EXPERTS              IS 1-3
EXPERT AVAILABILITY         IS over 90%
TURNOVER                    IS low
EXPERT ATTITUDE             IS willing

EXPERTISE                   IS AVAILABLE(100)
                               (Reliability = 100)

DATA CONTROL                IS moderate
PROCEDURE CONTROL           IS moderate
PERFORMANCE METRIC          IS high

CONTROL                     IS OK(100)
                               (Reliability = 100)

RISK                        IS LOW(33)
                               MODERATE(33)
                               HIGH(33)
                               (Reliability = 88)

SUITABILITY                 IS  FAIR(44)
                                GOOD(44)
                                POOR(11)
                                (Reliability = 58)
```

Figure 4-5. KBE Dialogue—Scheduler

3.2 Getting Started

Figure 4-5 is a dialogue between a user and TIMM that lets us follow TIMM's progress through our DFD of Figure 3-4. The user's input is in lower case; the system's responses are in upper case. The system actually begins by trying to apply the rules which are the mini-spec for Determine-Suitability as shown in Appendix B. However, it immediately comes up against a need to know the WORTH in order to determine Suitability and realizes that WORTH is the dataflow output from Determine-Worth.

At this point TIMM changes its focus to Determine-Worth and attempts to use the rules which are the mini-spec for Determine-Worth to find a value for WORTH, again as listed in Appendix B. Once again, it immediately discovers a need for information it does not have, specifically the Payoff/Cost ratio. It realizes that there is only one dataflow which can provide that information, and that must come from the user. Thus, the first interaction we see is a question to the user as shown in Figure 4-5:

```
PAYOFF/COST IS

to which the user has entered:

"over 3"
```

A similar situation applies to getting information about Percent-Solution and Type, but when those data are entered, the system has everything it needs to apply the Determine-Worth rules and infer a value for WORTH as shown on the next line of the listing.

3.3 Follow the Flow

Now having a value for the WORTH dataflow, the system shifts its focus back to Determine-Suitability. Once again it runs up against a need for further information in the rules, this time for Employee Acceptance, Solution Available, Easier Solution, and Teachability, which it must ask of the user.

Finally, the system comes to the need for a RISK dataflow, and in the process of determining RISK from the mini-spec rules of Determine-Risk, it invokes the rules for Determine-Complexity, Determine-Expertise, and Assess-Controls.

3.4 Reaching a Decision

When the system has some kind of input for each dataflow to a particular mini-system, it applies its inference strategy to the rules for that mini-system and develops the output dataflow. Although in these examples, each of the factors such as Payoff/Cost is stated by the user with 100% certainty, this is not necessary. One could say for example:

```
PAYOFF/COST IS          1.5 - 3 (50)
                        over 3 (50)
```

which indicates that the user is only 50 - 50 sure that Payoff/Cost is either 1.5 - 3 or over 3. In such a case, the system must incorporate this uncertainty into its inference strategy.

Furthermore, on each of the lines of Figure 4-5 on which a decision has been reached—WORTH IS or COMPLEXITY IS for examples—the answer may be one or more values, each followed by a number in parentheses and a Reliability estimate. The number in parentheses is TIMM's certainty of the decision being correct, and the Reliability is how sure it is that its decision certainty is correct, a number based on the similarity between the application we asked it to assess and the training rules in Appendix B. The more training cases it has, the higher its average reliability is likely to be.

In the remaining cases in this chapter, one can see the same processes at work in the dialogue listings.

4. Communications Network Diagnosis

A real-world communications network is a vast array of electronic switches and multiplexers distributed over a large geographical area, such as the United States. Its primary mission is to assure that a message entering the system at some location gets to its destination intact. Such networks may handle telephone messages, packets of data between computers, or even visual information.

Whatever the quality of information being sent and received, every network has error-detection hardware built into every node, and as errors occur they are transmitted as alarm signals to one or more control locations. Most of the millions of alarms per day are spurious, resulting from environmental noise, ripple effects of legitimate problems in other nodes, or even malfunction of the error circuitry.

In any case, the diagnostic problem is monumental, both in terms of its size and in terms of the tens of thousands of dollars a communications company may lose every minute that its network is down. In many companies, such as Generic, some highly qualified technicians seem to have a sixth sense about what alarms are real and how to locate and fix the problem. Too few of these people exist, however, and the tendency is to simply throw a barrage of technicians at the problem and hope somebody will solve it quickly.

Figure 4-6 shows the evaluation criteria for this application. The goal is to locate real problems in the network. One interesting aspect of solving an expert problem is that is frees up resources to tackle more complicated tasks. In this example, if there were fewer problems in identifying legitimate alarms, it might be possible to go another step and build an expert system to anticipate trouble in the network and reroute message traffic before a problem occurs.

```
Complexity-Factor = *Factors related to application complexity.*
      Decision-Definition      = *How well-defined is the decision?*
                                 [FUZZY; OK; WELL-DEFINED]
      Intuition/Common-Sense = *How much non-deductive thinking?*
                                 [<10%; 10%-50%; >50%]
      Knowledge-Domain         = *How broad is the experts' knowledge?*
                                 [ECLECTIC; OK; NARROW]
      Technology               = *What are the technological requirements?*
                                 [BUILD; ENHANCE; MODIFY; EXISTS]

Control-Factor    = *Factors related to administrative control.*
      Data-Control             = *Degree of control over data inputs*
                                 [LOW; MODERATE; HIGH]
      Performance-Metric       = *How good is the measure of expertise?*
                                 [LOW; MODERATE; HIGH]
      Procedure-Control        = *Degree of control over admin. procedures*
                                 [LOW; MODERATE; HIGH]

Expertise-Factor  = *Factors related to the domain experts*
      Expert-Attitude          = *What is the experts' attitude?*
                                 [HOSTILE;    INCOMPETENT;    UNINTERESTED;
                                 WILLING]
      Expert-Availability      = *For how much of the project will the*
                                 *experts be available?*
                                 [<50%; 50%-75%; 75%-90%; >90%]
      No.-of-Experts           = *How many domain experts are there?*
                                 [0; 1-3; >3]
      Turnover                 = *How high is the turnover of experts?*
                                 [LOW; MODERATE; HIGH]

Other-Factor      = *Factors not included elsewhere.*
      Employee-Acceptance      = *How will employees react to the system?*
                                 *This is a synonym for Culture.*
                                 [NEGATIVE; NEUTRAL; POSITIVE]
      Easier-Solution          = *Is there an easier way to solve the prob-
                                 lem?*
                                 [NONE; PARTIAL; COMPLETE]
      Solution-Available       = *How good is an existing solution?*
                                 [ADEQUATE; PARTIAL; NONE]
      Teachability             = *How easily is the skill taught?*
                                 [DIFFICULT; POSSIBLE; FREQUENT]

Worth-Factor      = *What is the value of the system?*
      Payoff/Cost              = *What is the estimated ratio of payoff to
                                 cost?*
                                 [<1; 1-1.5; 1.5-3; >3]
      Percent-Solution         = *How much of the problem will be solved*
                                 *by doing this system?*
                                 [<50%; 50%-75%; 75%-90%; >90%]
      Type                     = *Why is the application being considered?*
                                 [DEMO; USEFUL; NECESSARY; CRUCIAL]
```

Figure 4-6. KBE Evaluation Criteria—Communications Network

4.1 Technology

Diagnostic problems were some of the first applications of expert system thinking, and several systems have been created to perform diagnoses in medicine, equipment repair, and other areas. The technology for this application is readily available, and it is possible that some existing knowledge engineering tool could be used without modification. Almost the entire complexity of this application lies in the size of the knowledge base which, you may have noticed, was not included in our list of criteria. It's likely that in another refinement of KBE we might add a factor to deal with how long it takes a human to do the task as a measure of its complexity.

On the other hand, it is likely that we would like to hook up the expert system to the network itself so that alarm signals can be detected as they occur. For this reason, I have chosen the criterion which suggests we will probably have to modify some existing system. The fact that some major hardware changes may also be needed to hook the system into the network is not really an issue, since this work would be necessary whether or not knowledge-based technology was used to solve the problem.

4.2 Expert and Employee Attitude

I have chosen WILLING as the value for Expert Attitude, but this is the kind of situation where experts often start out uninterested because they don't believe the job can be done. Some technicians, particularly those who are least expert, may even be hostile toward the development.

If the early prototype knowledge base shows promise, the real experts' attitude will probably change to one of willingness as they realize how the system can help on the job. In terms of knowledge base development, the hostile quasi-experts will probably not be involved and hence not a threat to the project. Also, the fairly high turnover of technicians makes it likely that many of this group of people will leave before being replaced by the system.

If this system is successful, many of those non-experts could be replaced. If they have not left on their own, then the existence of the system could have a negative effect on employee attitude. For this example, I have chosen a NEUTRAL value to describe employee acceptance.

4.3 Suitability

The other choices are more or less self-explanatory, and almost all suggest that this is an ideal application. We would expect KBE to tell us that it is a GOOD one. The early expert system technology evolved around diagnostic problems such as the MYCIN medical diagnosis system. At this time, nearly any true diagnostic problem is a good application of expert system technology if there are measurable data which humans use deductively to reach a diagnosis whose correctness can be determined.

The KBE evaluation of the suitability of this application is shown in Figure 4-7.

5. Microelectronic Sales

Generic, like many companies, often has to send relatively inexperienced negotiators into battle with customers who really know the ropes in their area. The question is whether or not expert systems can help.

The problem really is to analyze the personality and tactics of the customer's negotiator as compared with Generic's expert negotiator and to provide extensive, specific information and guidelines for Generic salespeople to use in negotiating sessions. This application is a tricky candidate in that at the outset it appears to have badly defined output, a potentially vast knowledge base, little possibility of administrative control, and various other shortcomings. Perhaps as we look more closely at the selection criteria in Figure 4-8, we will find some useful subset of a complete solution.

5.1 Decision Definition

It's not at all clear what the output of this system should be. It might offer qualitative guidelines such as:

```
FOR THIS PERSON, TRY FLATTERY, or
THIS PERSON'S FIRST OFFER IS USUALLY HER LAST OFFER
```

Or it might offer more quantitative guidelines such as:

```
20% IS THE MAXIMUM DISCOUNT FOR THIS CUSTOMER
```

For these types of output we definitely need an expert's evaluation of the customer in question, but an inference engine seems almost superfluous. It would be almost as useful to simply list the guidelines which have been accumulated for any given negotiator.

What's needed is a system which understands the negotiating environment in-depth and can use characteristics of individual negotiators to become a participant in the negotiation. This is a situation in which the rule-based knowledge engineering tools now available begin to fall short. A more powerful representation scheme is needed, such as the script-based technology described in Chapter 9, The Knowledge-Based Information Center. The output is definintely fuzzy, but script-based systems can often handle such situations if the negotiation script can be described.

```
***********************************************************************
        © Copyright 1984 General Research Corporation
          TIMM (TM) The Intelligent Machine Model
                        Version 2.0
***********************************************************************

        PAYOFF/COST                  IS over 3
        PERCENT SOLUTION             IS over 90%
        TYPE                         IS necessary

        WORTH                        IS HIGH(100)
                                        (Reliability = 100)

        EMPLOYEE ACCEPTANCE          IS neutral
        SOLUTION AVAILABLE           IS none
        EASIER SOLUTION              IS none
        TEACHABILITY                 IS frequent

        INTUITION/COMMON SENSE       IS under 10%
        TECHNOLOGY                   IS modify
        DECISION DEFINITION          IS well-defined
        KNOWLEDGE DOMAIN             IS narrow

        COMPLEXITY                   IS LOW(100)
                                        (Reliability = 78)

        NO. OF EXPERTS               IS 1-3
        EXPERT AVAILABILITY          IS over 90%
        TURNOVER                     IS high
        EXPERT ATTITUDE              IS willing

        EXPERTISE                    IS AVAILABLE(100)
                                        (Reliability = 100)

        DATA CONTROL                 IS high
        PROCEDURE CONTROL            IS high
        PERFORMANCE METRIC           IS high

        CONTROL                      IS TIGHT(100)
                                        (Reliability = 100)

        RISK                         IS LOW(100)
                                        (Reliability = 100)

        SUITABILITY                  IS GOOD(100)
                                        (Reliability = 56)
```

Figure 4-7. KBE Dialogue—Communications Network

```
Complexity-Factor = *Factors related to application complexity.*
      Decision-Definition      = *How well-defined is the decision?*
                                 [FUZZY; OK; WELL-DEFINED]
      Intuition/Common-Sense   = *How much non-deductive thinking?*
                                 [<10%; 10%-50%; >50%]
      Knowledge-Domain         = *How broad is the experts' knowledge?*
                                 [ECLECTIC; OK; NARROW]
      Technology               = *What are the technological requirements?*
                                 [BUILD; ENHANCE; MODIFY; EXISTS]

Control-Factor    = *Factors related to administrative control.*
      Data-Control             = *Degree of control over data inputs*
                                 [LOW; MODERATE; HIGH]
      Performance-Metric       = *How good is the measure of expertise?*
                                 [LOW; MODERATE; HIGH]
      Procedure-Control        = *Degree of control over admin. procedures*
                                 [LOW; MODERATE; HIGH]

Expertise-Factor  = *Factors related to the domain experts*
      Expert-Attitude          = *What is the experts' attitude?*
                                 [HOSTILE;   INCOMPETENT;   UNINTERESTED;
                                 WILLING]
      Expert-Availability      = *For how much of the project will the*
                                 *experts be available?*
                                 [<50%; 50%-75%; 75%-90%; >90%]
      No.-of-Experts           = *How many domain experts are there?*
                                 [0; 1-3; >3]
      Turnover                 = *How high is the turnover of experts?*
                                 [LOW; MODERATE; HIGH]

Other-Factor      = *Factors not included elsewhere.*
      Employee-Acceptance      = *How will employees react to the system?*
                                 *This is a synonym for Culture.*
                                 [NEGATIVE; NEUTRAL; POSITIVE]
      Easier-Solution          = *Is there an easier way to solve the prob-
                                 lem?*
                                 [NONE; PARTIAL; COMPLETE]
      Solution-Available       = *How good is an existing solution?*
                                 [ADEQUATE; PARTIAL; NONE]
      Teachability             = *How easily is the skill taught?*
                                 [DIFFICULT; POSSIBLE; FREQUENT]

Worth-Factor      = *What is the value of the system?*
      Payoff/Cost              = *What is the estimated ratio of payoff to
                                 cost?*
                                 [<1; 1-1.5; 1.5-3; >3]
      Percent-Solution         = *How much of the problem will be solved* *by
                                 doing this system?*
                                 [<50%; 50%-75%; 75%-90%; >90%]
      Type                     = *Why is the application being considered?*
                                 [DEMO; USEFUL; NECESSARY; CRUCIAL]
```

Figure 4-8. KBE Evaluation Criteria—Microelectronics Sales

So, the ideal would be that when the customer says something, we input that to our expert negotiating system, which then tells the Generic negotiator what to say next.

5.2 Intuition/Common Sense

Rules of negotiation are actually fairly rigorous, although they differ somewhat from person to person. The crucial difference between an expert negotiator and one who is not expert is their knowledge of when to apply what strategies. This partakes heavily of non-deductive thinking, in this case probably more than half.

5.3 Technology

The script-based technology[2] is needed to approach a satisfactory solution to this problem. Although several expert systems have been developed for marketing, accounting, and similar applications, finding an off-the-shelf system to do this job is unlikely.

Not only that, but having inexperienced in-house knowledge engineers or programmers do the job seems a poor choice. In-house experts would probably have to work with a vendor to make enhancements to existing technology. These enhancements are so extensive, however, that I have chosen the BUILD value for the technology criterion.

5.4 Easier Solution

The expertise for this job appears to be available, but there doesn't seem to be an easy match with current technology. It may be better at this time to simply have the expert write down known characteristics and idiosyncracies of the customers' people as best he knows and provide that list to our negotiators as needed.

Human Edge Software has taken an approach something like this with their systems called *Sales Edge* and *Negotiating Edge.* These are not knowledge-based systems in the current sense of the term, but some people feel they do a useful job. The technique is based on using psychological inventories of personality characteristics of both parties and simply providing guidelines based on those characteristics. The systems run on an IBM PC and offer a cheap partial solution to our problem.

5.5 Suitability

I feel that this is a POOR application to try early in a company's experience with expert systems, primarily because of the high risk engendered by the complexity of the application. Perhaps a negotiating system could be tackled after developing a more controlled script-based application, such as accounting. Even then, however, a clearly HIGH Payoff/Cost ratio is a prerequisite for an application of this complexity.

[2] Cognitive Systems, Inc. of New Haven, Conn. has the most highly developed understanding of this technology.

```
*****************************************************************
      © Copyright 1984 General Research Corporation
         TIMM (TM) The Intelligent Machine Model
                      Version 2.0
*****************************************************************

PAYOFF/COST                  IS 1.5-3
PERCENT SOLUTION             IS 75%-90%
TYPE                         IS necessary

WORTH                        IS HIGH(100)
                                (Reliability = 63)

EMPLOYEE ACCEPTANCE          IS positive
SOLUTION AVAILABLE           IS none
EASIER SOLUTION              IS partial
TEACHABILITY                 IS possible

INTUITION/COMMON SENSE IS over 50%
TECHNOLOGY                   IS build
DECISION DEFINITION          IS fuzzy
KNOWLEDGE DOMAIN             IS ok
COMPLEXITY                   IS HIGH(100)
                                (Reliability = 100)

NO. OF EXPERTS               IS 1-3
EXPERT AVAILABILITY          IS over 90%
TURNOVER                     IS moderate
EXPERT ATTITUDE              IS willing

EXPERTISE                    IS AVAILABLE(100)
                                (Reliability = 100)

DATA CONTROL                 IS low
PROCEDURE CONTROL            IS low
PERFORMANCE METRIC           IS high

CONTROL                      IS LOOSE(100)
                                (Reliability = 100)

RISK                         IS HIGH(100)
                                (Reliability = 100)

SUITABILITY                  IS POOR(100)
                                (Reliability = 59)
```

Figure 4-9. KBE Dialogue—Microelectronics Sales

The KBE solution is shown as Figure 4-9.

6. Insurance

Many candidate applications for expert system technology can be found in the insurance industry, including underwriting, policy pricing, and product information systems. For this example we will look at the underwriting area. Figure 4-10 shows the evaluation criteria.

Many of the choices in this situation are self-explanatory; fundamentally, a simple yes/no decision is being made about underwriting the risk of insuring a particular person, although certain premium issues may enter the picture. The expertise is readily available and often already codified, and both the company and employee are likely to benefit from the system.

In addition, many of the currently available knowledge engineering tools can handle this application. Modification may be necessary in their input/output facilities and their ability to get data from files instead of from the user only, but these are principally data processing issues.

```
Complexity-Factor             = *Factors related to application complexi-
                                ty.*
      Decision-Definition     = *How well-defined is the decision?*
                                [FUZZY; OK; WELL-DEFINED]
      Intuition/Common-Sense  = *How much non-deductive thinking?*
                                [<10%; 10%-50%; >50%]
      Knowledge-Domain        = *How broad is the experts' knowledge?*
                                [ECLECTIC; OK; NARROW]
      Technology              = *What are the technological requirements?
                                [BUILD; ENHANCE; MODIFY; EXISTS]

Control-Factor                = *Factors related to administrative con-
                                trol.*
      Data-Control            = *Degree of control over data inputs*
                                [LOW; MODERATE; HIGH]
      Performance-Metric      = *How good is the measure of expertise?*
                                [LOW; MODERATE; HIGH]
      Procedure-Control       = *Degree of control over admin. procedures
                                [LOW; MODERATE; HIGH]

Expertise-Factor              = *Factors related to the domain experts*
      Expert-Attitude         = *What is the experts' attitude?*
                                [HOSTILE;   INCOMPETENT;   UNINTERESTED
                                WILLING]
      Expert-Availability     = *For how much of the project will the*
                                *experts be available?*
                                [<50%; 50%-75%; 75%-90%; >90%]
      No.-of-Experts          = *How many domain experts are there?*
                                [0; 1-3; >3]
```

```
     Turnover                  = *How high is the turnover of experts?*
                                 [LOW; MODERATE; HIGH]

Other-Factor                   = *Factors not included elsewhere.*
     Employee-Acceptance       = *How will employees react to the system?*
                                 *This is a synonym for Culture.*
                                 [NEGATIVE; NEUTRAL; POSITIVE]
     Easier-Solution           = *Is there an easier way to solve the prob-
                                 lem?*
                                 [NONE; PARTIAL; COMPLETE]
     Solution-Available        = *How good is an existing solution?*
                                 [ADEQUATE; PARTIAL; NONE]
     Teachability              = *How easily is the skill taught?*
                                 [DIFFICULT; POSSIBLE; FREQUENT]

Worth-Factor                   = *What is the value of the system?*
     Payoff/Cost               = *What is the estimated ratio of payoff to
                                 cost?*
                                 [<1; 1-1.5; 1.5-3; >3]
     Percent-Solution          = *How much of the problem will be solved*
                                 *by doing this system?*
                                 [<50%; 50%-75%; 75%-90%; >90%]
     Type                      = *Why is the application being considered?*
                                 [DEMO; USEFUL; NECESSARY; CRUCIAL]
```

Figure 4-10 KBE Evaluation Criteria—Insurance Underwriting

6.1 No. of Experts

From the insurance company's point of view, there is only one source of expertise: not underwriters, but the actuarial rules that underwriters use. The assumption is that underwriters apply these rules as they are written. Although I know of no rigorous study of this situation, it seems likely that underwriters sometimes apply the rules creatively where they feel the rules don't quite fit an existing situation.

Nonetheless, an excellent knowledge base can be built simply by structuring the actuarial rules for use with some knowledge engineering tool.

6.2 Payoff/Cost

The dollar payoff of such a system is very difficult to assess, since the underwriting job being done now is generally satisfactory. The real payoff for the insurance company is the release of underwriters from a relatively mundane task so their talents can be used elsewhere in the company. The dollar payoff then is in reassigning the underwriting staff to jobs that would otherwise require new employees.

6.3 Suitability

As shown in Figure 4-11, KBE agrees that this is a GOOD application because of its relatively low complexity and potentially high Payoff/Cost ratio. However, because most insurance companies already have an adequate underwriting system, only the most forward-looking of them will attempt to automate it.

7. Conclusion

The mix of applications presented here barely scratches the surface of candidates for expert systems. The structure for KBE is useful as it is, but in practice it may require some modifications. Although these should be made cautiously, with an awareness of the impact on other criteria, it is in the nature of knowledge bases to grow. KBE is no exception.

Although we discussed only a few applications, the issues raised and the thought process involved in these cases are common to the evaluation of a much broader range of candidates. The application selection work done in the Survey is among the most important phases of an expert system project. Careful selection here can save a boatload of troubles later.

In the next chapter, we look more deeply at the seminar scheduling application as we enter the structured analysis phase of the project.

```
*******************************************************************************
          © Copyright 1984 General Research Corporation
             TIMM (TM) The Intelligent Machine Model
                            Version 2.0
*******************************************************************************

        PAYOFF/COST                 IS over 3
        PERCENT SOLUTION            IS over 90%
        TYPE                        IS useful

        WORTH                       IS HIGH(100)
                                       (Reliability = 100)

        EMPLOYEE ACCEPTANCE         IS neutral
        SOLUTION AVAILABLE          IS none
        EASIER SOLUTION             IS none
        TEACHABILITY                IS frequent

        INTUITION/COMMON SENSE      IS under 10%
        TECHNOLOGY                  IS modify
        DECISION DEFINITION         IS well-defined
        KNOWLEDGE DOMAIN            IS narrow

        COMPLEXITY                  IS LOW(100)
                                       (Reliability = 78)

        NO. OF EXPERTS              IS 1-3
        EXPERT AVAILABILITY         IS over 90%
        TURNOVER                    IS high
        EXPERT ATTITUDE             IS willing

        EXPERTISE                   IS AVAILABLE(100)
                                       (Reliability = 100)

        DATA CONTROL                IS moderate
        PROCEDURE CONTROL           IS moderate
        PERFORMANCE METRIC          IS high

        CONTROL                     IS OK(100)
                                       (Reliability = 100)

        RISK                        IS LOW(100)
                                       (Reliability = 100)

        SUITABILITY                 IS GOOD(100)
                                       (Reliability = 56)
```

Figure 4-11. KBE Dialogue—Insurance Underwriting

Chapter 5
A Case Study in Structured Analysis

In the last chapter, we evaluated several potential AI applications and chose the scheduling of Generic Enterprises' commercial seminars as the application to be followed into analysis. Before beginning the work of analyzing the expert Scheduler, however, we need to discuss further the purpose and goals of analysis, as well as the general structure of expert systems.

As we proceed through this discussion it should become clear that analysis is the single most important phase of any development project. This is particularly true for knowledge-based systems since the analysis phase includes the primary structural definition of the knowledge base. If analysis is done thoroughly and with integrity, then the rest of the project is likely to be very straightforward.

1. Goals of Analysis

Analysis means examining a system and identifying its component pieces while preserving the relationship among those pieces so that an equivalent system can be synthesized. This is exactly our mission in doing analysis for the purpose of building an expert system: We wish to discover the functional pieces that go into the human expert's performance so that a machine can be programmed to do those same tasks in approximately the same way the human does them.

Analysis also has come to mean creating a model of the system that eventually will be in production. The idea of modeling is certainly not new—it's been used by architects and other designers for a long time. The idea of modeling software systems is relatively recent, however, having gained popularity in the 1970s through the use of techniques such as DFDs for modeling in a structured analysis.

2. Why Structured Analysis?

We have already learned quite a bit about structured analysis through discussions of the analysis process in Chapter 1 and the DFD and data dictionary tools in Chapter 3. Now we want to look at some good reasons why we should prefer the structured approach over other analytical techniques. All of them have to do with communication: clarity, project management, and user interaction, to name a few.

2.1 Clarity of Specification

Historically, system development projects get in trouble for one major reason: a failure of communication. The failure may occur between user and analyst over the goals of the project, between analyst and designer or programmer over what's to be built, or between management and everybody over project team interaction. More often than not these troubles result from lack of a suitable communication medium. The tools and techniques of structured analysis create a bridge to cross these communication gaps.

The traditional system specification is a lengthy narrative document, too often one in which process and data descriptions are ambiguous, described twice or three times in different parts of the spec, and incomplete.[1] The DFD technique of structured analysis makes it clear that a picture is worth a thousand words since on a few pages of DFD all of a system's functions and data interfaces are represented in easy-to-understand packages.

2.2 Project Management

From a project management point of view, structured analysis provides some important benefits. Primary among these is the ability to view the system graphically in a top-down fashion. Further, and particularly relevant to expert systems, a structured analysis enables the developer to iterate over the parts of an application until they achieve satisfactory performance. This stands in sharp contrast to the more traditional approach of having fixed points in the project life cycle in which working documents become frozen deliverables.

[1] The important criteria of any successful specification are that it be complete, unambiguous, and non-redundant.

2.3 User Interaction

Another important feature of the structured approach is its insistence on a high degree of interaction between user and analyst. As we saw in discussing the selection of expert system applications, it is crucial that user and analyst work hand in hand over a long period of time to evolve the knowledge base to a mature state.

3. Expert System Structure Revisited

In Chapter 3 we drew a DFD, Figure 3-4, to represent the KBE system and represent the structure of its **knowledge base.** In the case of KBE, the rules represented the expertise used in drawing conclusions about the suitability of an application for knowledge-based system treatment.

We saw also that we may have to draw DFDs, such as Figures 3-6 and 3-7, for the part of the system that knows how to reach conclusions using the rules in the knowledge base: the **inference engine.** The inference engine is much like a child who has the inherent ability to use its experience to make reasonable decisions, but has little or no experience about which to reason. In order to understand the remainder of this chapter and those that follow, we need to review our previous thoughts about the nature of the inference engine and how it relates to the knowledge base.

Every computer system, however traditional, has an inference engine—that is, reasoning abilities. In a conventional COBOL program it is represented by those parts of the program code which control access to the right well-defined business data at the right time and lead in some well-determined way to well-determined logic statements, which produce outputs appropriate to the business problem being solved.

A conventional program typically has a very limited inference capability that is highly intertwined with the knowledge it has to process. In addition, a conventional system's knowledge typically consists mainly of the facts in its database along with well-determined, built-in relationships among those facts. When we draw DFDs for a conventional system, we never see the inference capability as a separate entity because it is almost always a simple sequential execution of program statements as they are encountered.

An expert system inference engine is a not-always-well-defined program which accesses not only data, but also pieces of the program—the knowledge base rules—and is able to integrate and use them in ways which may not be completely predictable ahead of time, that is to say heuristically. In my view, one of AI research's giant leaps forward is the realization that knowledge and the software that processes it are separate entities, entities which can be built, used, and maintained quite independently of one another. As a result, several techniques have evolved to implement reasoning explicitly. One of these, rule chaining, is discussed further in Chapter 7, Programming Expert Intelligence.

Because the knowledge base consists of executable rules—the pieces of program used by an inference engine—we find ourselves having to use DFDs to represent its structure and content as we did in Chapter 3. If a particular application has multiple expert mini-systems with inference engines which differ from each other or from any that have been written previously, we may find ourselves having to draw DFDs to represent their reasoning logic as well.

This split personality in an expert mini-system often forces us to draw an additional level of DFD which shows the analysis of the inference engine and its specific interrelationships with a knowledge base. In many instances, fortunately, typically when we use a domain-independent inference engine such as a commercial expert system shell, this additional detail is not necessary. This was the case in previous chapters when we specified the KBE and assumed TIMM to be its universally applied inference engine.

In the analysis of the Scheduler system which follows, we are not so lucky. In fact we find several domain-specific expert mini-systems for which both inference engine and knowledge base must be specified. (In Chapter 6, Structured Analysis and PROLOG, we will switch back to analyzing the knowledge base exclusively since the programming language PROLOG is itself an inference engine which will apply throughout the chapter.)

4. Why Scheduling?

Admittedly, the Scheduler is a relatively simple, though not a trivial, application of AI technology. One minor reason for choosing it is that our KBE system said in the last chapter that it's an OK application.

Another reason is that the Scheduler is a project of manageable size, not only in the context of this book, but also in the real world. For some reason there is a rumor going around that a legitimate AI application must cost at least $500,000 and more likely $1-3 million; I regard this as absurd. Knowledge-based techniques can be applied in relatively small ways to substantially improve many traditional applications, and I would recommend that for a first project at least, a company should choose an application which can attain reasonably expert performance in a year or two at most.

Yet another, perhaps more cogent, reason for choosing this application is that it clearly illustrates my assertion of the balance of importance of the AI component in a particular system relative to other technologies.[2] While the proportion may vary from application to application, knowledge-based processing is hardly ever the only major system component. In other words, there is still plenty of room for traditional analysis, design, and programming skills, not to mention well-known information modeling techniques.

Finally, I choose the Scheduler application both because the AI component is not restricted to just one part of the system, and because there is a need both for structured decision making and planning expertise, the two major types of expert system.

[2] Reid G. Smith of Schlumberger-Doll Research further supports this thought by observing that in the Dipmeter Advisor system the inference engine and knowledge base accounted for only 30% of the system code. The remainder was used for feature detection, user interface, and support environment functions. See Smith, R.G., "On the Development of Commercial Expert Systems," *The AI Magazine*, Fall 1984, p. 65.

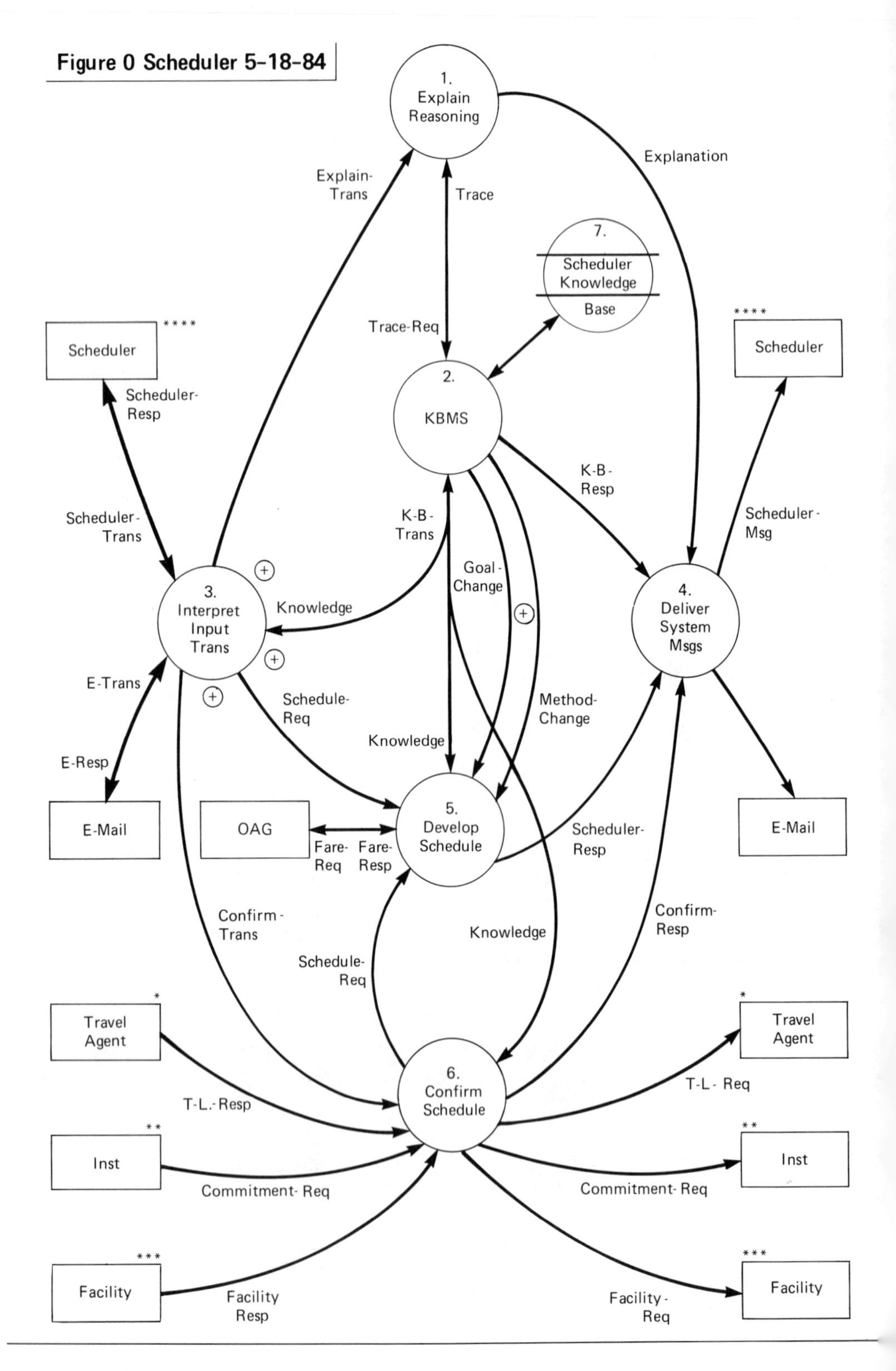

Figure 5-1. Scheduler Overview

5. DFD Figure 0 - The Expert Scheduler Overview

We are now ready to look more closely at the knowledge-based Scheduler system. My purpose in this chapter is not to illustrate all of the steps in a structured analysis; this is done quite adequately in other texts.[3] Neither do I intend to provide a complete specification of the system.

Rather, I wish to illustrate the use of structured techniques in specifying a system in which some, but not all, of the components are knowledge-based, discussing particularly the blend of new and old technology. As we walk through the data flow diagrams, it's important to be aware of where we meet AI and where we are dealing with traditional methodologies; the difference is not always clear.

Figure 5-1 shows a high-level DFD of the entire Scheduler system, a representation of the way we want Generic's system to emulate the human scheduler's work. Referring to the phases of structured analysis described in Chapter 1, this is roughly equivalent to the new logical model of the system. That is, we presume to have previously documented the way the human scheduler now does her job, removed from that documentation all traces of technology she currently has to deal with, and added in whatever new functional requirements we wish to have in the expert system. The conventions of structured analysis constrain us to number this DFD **Figure 0,** and to name it for the system it represents: **Scheduler.**

As with most DFDs, the easiest way to understand the diagram is to walk through the general processing of each transaction that can enter the system and then study the details of the processes it passes through. Only two major input transactions drive the system—Scheduler-Trans and E-Trans—and they are very similar. The other inputs—T-L-Resp, Fare-Resp, Commitment-Resp, and Facility-Resp—are auxiliary inputs used by some of the system's major functions.

Each of the major inputs consists of several types of detail transactions as shown by their definitions in the partial data dictionary of Figure 5-2. We will deal mainly with the details of Scheduler-Trans, the processing of which is illustrated with heavy lines overlaid on Figure 5-1 and shown as Figures 5-1a through 5-1d.

```
Commitment-Req   = *Request to instructor for an engagement*
                 = Start-Date + End-Date + Location + Engagement
Commitment-Resp  = *Response from Instructor to Commitment-Req*
                 = Date + Engagement + Disposition
Confirm-Trans    = *Request to commit to a schedule*
                 = CONFIRM + Schedule-#
Disposition      = *Instructor's commitment to an engagement*
                 = [NEGATIVE; HOLD + Hold-Date; POSITIVE]
E-Trans          = *Logical input from E-Mail system*
```

[3] See Keller (1983), op. cit.; DeMarco, T., Structured Analysis and System Specification, Yourdon Press, N.Y., 1978; McMenamin, S.M. and Palmer, J.F., Essential Systems Analysis, Yourdon Press, N.Y., 1984

```
                  = [Schedule-Req; Explain-Trans; E-KB-Trans]
E-KB-Trans        = *E-Mail request to the KBMS*
                  = [KB + KB-Query]
Explain-Trans     = *Request to explain reasoning*
                  = EXPLAIN + Schedule-# + ([SUITABILITY; GUID-
                    ANCE])
Facility-Resp     = *Response from Facility to Facility-Req*
                  = Facility + Date + Disposition
Fare-Resp         = *Response from OAG to Fare-Req*
                  = Date + Origin + Dest + {Airline + Flight + Fare}
KB-Trans          = *Request to the KBMS*
                  = [KB + Rule-Update; Check-Trans; KB-Query;
                    Frame-Update]
Schedule-Req      = *Request to develop alternative schedules*
                  = Subset-Monday + Subset-Friday
Scheduler-Trans   = *Logical input from the scheduler*
                  = [Schedule-Req; Explain-Trans; KB-Trans; Con-
                    firm-Trans]
T-L-Resp          = *Response from travel agent to T-L-Req*
                  = Instructor + Date + {Origin + Dest + Airline +
                    Flight + Time + Status}
```

Figure 5-2. Scheduler—Partial Data Dictionary

5.1 Scheduler-Trans

The human scheduler is still the primary driving force in the operation of the system. She is first of all, solely responsible for operational interaction with the knowledge base (KB-Trans), whether it is to modify rules and data or check consistency. In addition, she has access to the Knowledge Base Management System's (KBMS's) explanation facility (Explain-Trans), starts the automatic scheduling process (Scheduler-Trans), and is the one who commits to a particular schedule (Confirm-Trans). Further, she has available a generalized query facility to get information about any of the data or knowledge in the knowledge base. In this relationship, the expert Scheduler is an extension of the intelligence of the human scheduler, subsequently called the system user.

5.1.1 Interpret-Input-Trans

Each of these transactions is entered into Process 3,[4] Interpret-Input-Trans whose job is to translate them into a standard form and switch each of them to the appropriate functional processor.

[4] The numbering of processes on DFDs is for identification purposes only and carries no significance as to the order in which processes are used.

Interpret-Input-Trans is one of those places where we may or may not find knowledge-based processing. It depends partly, for example, on what kind of query language the user can use. If it is a simple menu choice facility, this is probably a fairly simple process not requiring AI at all. On the other hand, if a natural English query facility is desired, then a fairly sophisticated AI processor will be required. (The natural English alternative is explored in Section 9 of this chapter, where we examine DFD Figure 3, the details of Interpret-Input-Trans.)

At this stage of the specification process, we are not necessarily committed to doing Interpret-Input-Trans one way or the other. In fact, depending on our implementation priorities, we may first choose to build Process 3 as a very simple processor and later add more sophistication.

5.1.2 KB-Trans

To get a feel for the information handled by Scheduler and the flexibility available to the user, let's deal first with the KB-Trans part of Scheduler-Trans shown in Figure 5-1a.

As KB-Trans passes through Interpret-Input-Trans, it is sent to Process 2, the Knowledge Base Management System, or KBMS. The tasks undertaken by a KBMS are very much like those handled by a traditional database manager (DBMS),[5] except that the KBMS must deal not only with data, but also with what we are calling knowledge in the form of rules, which are stored in a data structure called a **frame.**[6] The KBMS is the sole source of interaction with the Scheduler-Knowledge-Base, which is a highly structured repository for all of the data and knowledge in the system.

If a KB-Trans is entered that changes either the rules for preparing schedules or data affecting schedules, such as an instructor's availability, the KBMS sends a Goal-Change or Method-Change to Process 5, Develop-Schedule.[7] Based on such changes, Develop-Schedule may recreate a set of tentative schedules which subsequently can be displayed for the user. Develop-Schedule contains much of the inference technology in the system and can be thought of as the central AI component. As we have seen, however, other processes may also contain some heuristic processing.

When the KB-Trans has been processed by the KBMS, a KB-Resp is sent to another non-AI process, Process 4, Deliver-System-Msgs. Its job is simply to format and route system responses to the appropriate user, in this case the human scheduler.

5.1.3 Schedule-Req

Developing schedules is a major part of the raison d'etre of this system, and it is the Schedule-Req variety of Scheduler-Trans that makes it happen, as shown by Figure 5-1b.

[5] Although DBMSs vary considerably in their capabilities, their general mission is to be a relatively transparent link between the informational needs of an end-user or programmer and a large assemblage of physical data files. Basic functions of a DBMS usually include at least data modification and retrieval.

[6] Refer to Chapters 7 and 8 for a more complete discussion of knowledge representation techniques.

[7] Develop-Schedule is examined in more detail when we look at DFD Figure 5 in Section 6 of this chapter.

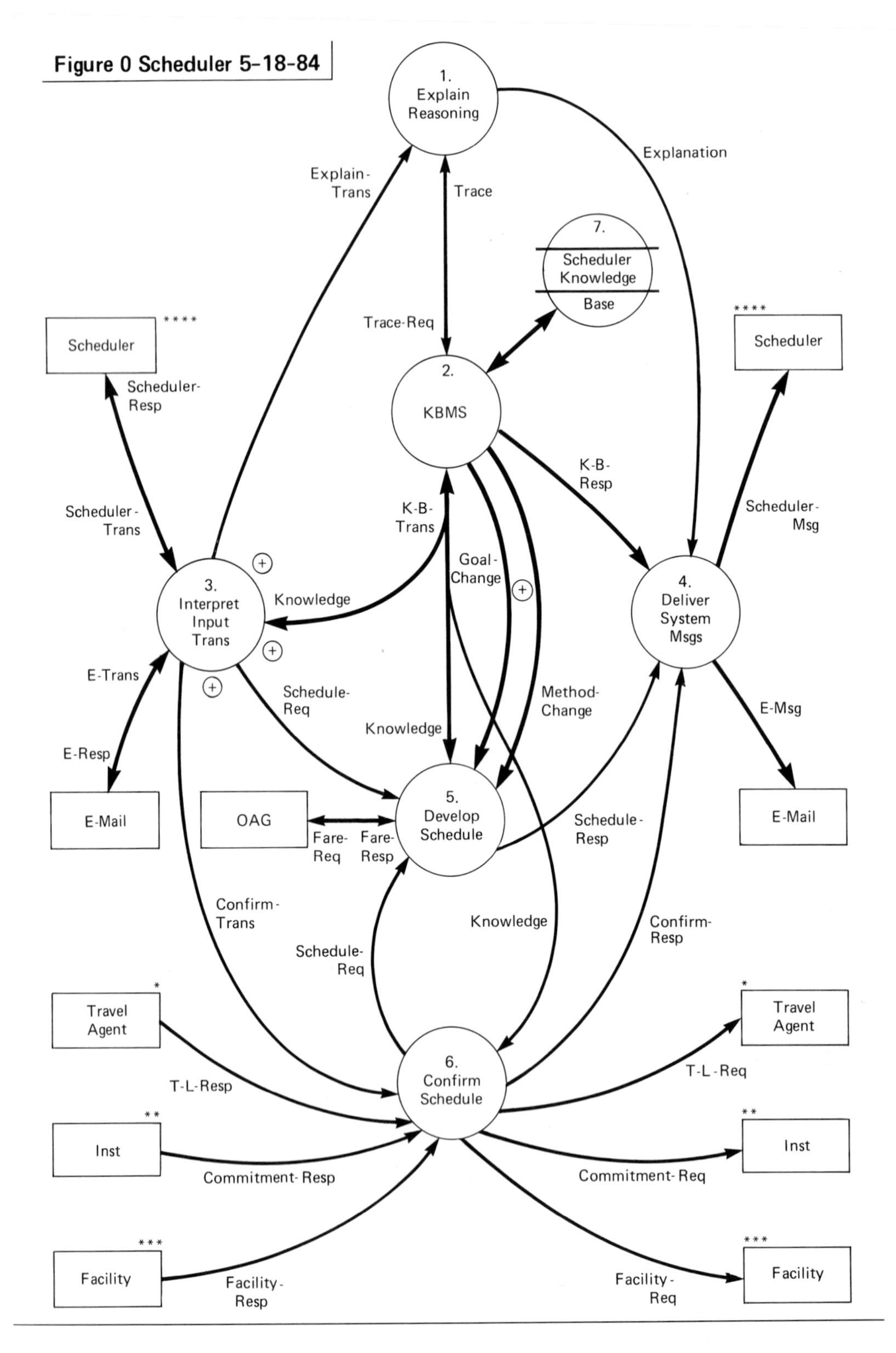

Figure 5-1a. KB-Trans Processing

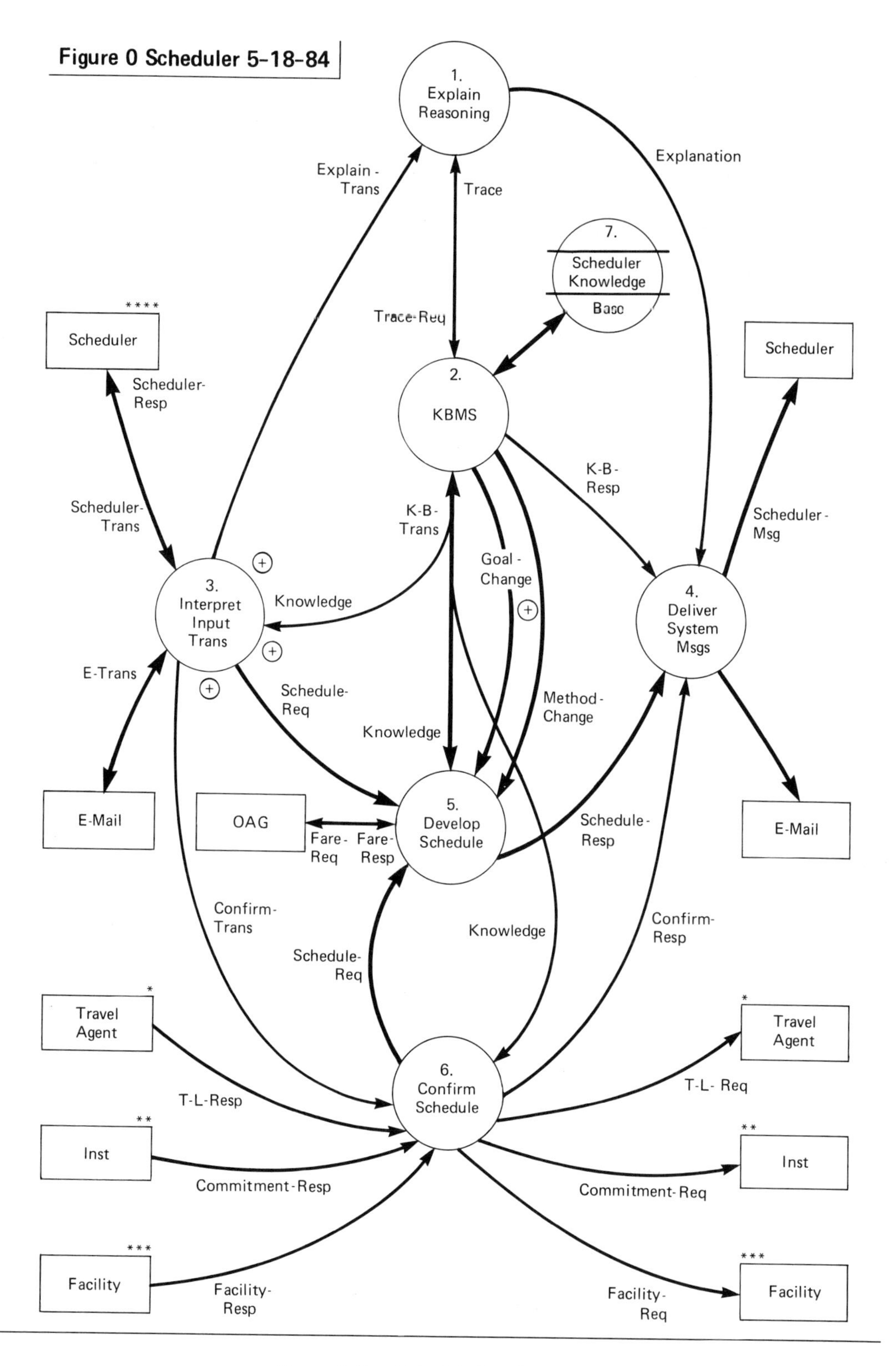

Figure 5-1b. Schedule-Req Processing

There is always just one current, though very dynamic, schedule in the system which Confirm-Schedule is always trying to implement, and which Develop-Schedule generally uses as a starting point for developing its alternative schedules. In doing this it considers such factors as instructor availability, de facto assignments, committed resources, and many others.

Develop-Schedule uses the current schedule along with heuristic rules which it gets via the KBMS from the knowledge base to develop several alternative schedules which meet its evaluation criteria. When a satisfactory set of alternatives has been developed, they are sent to Deliver-System-Msgs and then to the user for final judgment.

Please note that however expert this Scheduler system may be, the human remains an integral part of the decision process as the final judge of the best schedule. This Scheduler system might better be called The Scheduler's Apprentice. It may be that over a period of weeks, months, or years of production usage, the system can develop an expertise which, through experience, we find to be reliable enough to replace human experts. However, I stress once again that it is generally folly to begin an expert system project with the idea that some number of people are going to be replaced in a short period of time.

5.1.4 Explain-Trans

An inference engine's ability to make credible decisions in the face of uncertain data and conclusions is one of its major assets. This ability is also responsible for one of the main difficulties in evolving a knowledge base: knowing what rules were used to reach a particular conclusion.

For this reason, one of the most important features of any expert system is a facility to explain the reasoning it used to reach a conclusion. Figure 5-1c shows that in our system, the user enters an Explain-Trans which is sent by Interpret-Input-Trans to Process 1, Explain-Reasoning. It uses the KBMS which accesses the rules and data for the last schedule developed to trace the rules which were invoked in developing the schedule. It sends this information back to the user by way of Deliver-System-Msgs.

5.1.5 Confirm-Trans

Once the user has an acceptable schedule from the system, she enters the Confirm-Trans variety of Scheduler-Trans, which is sent by Interpret-Input-Trans to Process 6, Confirm-Schedule. (We will examine the details of Confirm-Schedule in Section 8 of this chapter.) This processing is shown by Figure 5-1d.

The job of Confirm-Trans is to access the details of a particular schedule and to manage the resources needed to implement the schedule. This involves interacting with instructors, travel agents, and facilities such as hotels where engagements may be held. Confirm-Schedule keeps track of these arrangements and makes appropriate adjustments when changes occur in the schedule. If it becomes impossible to implement the schedule as presented, Confirm-Schedule modifies the knowledge base to reflect the conflict and

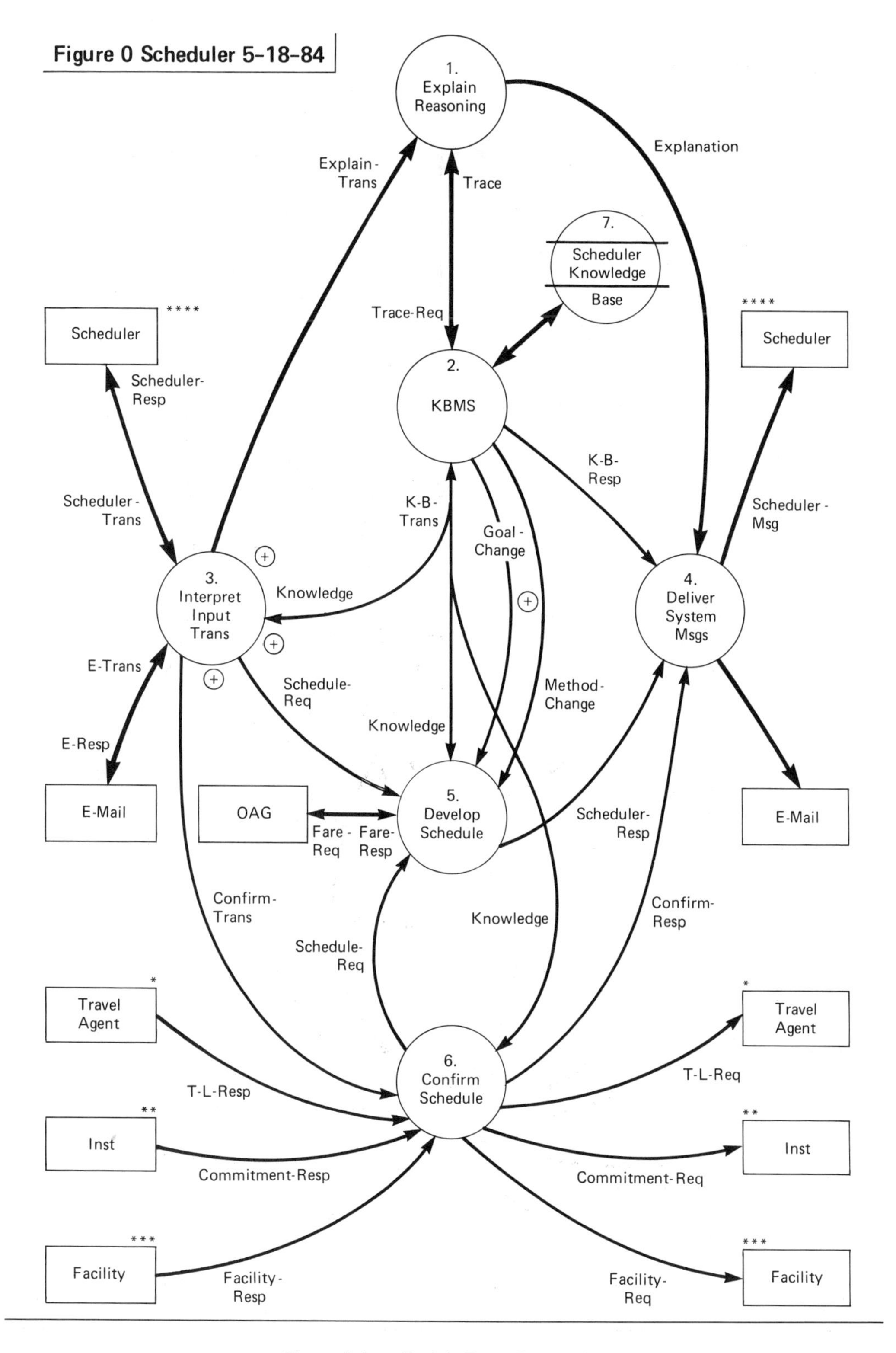

Figure 5-1c. Explain-Trans Processing

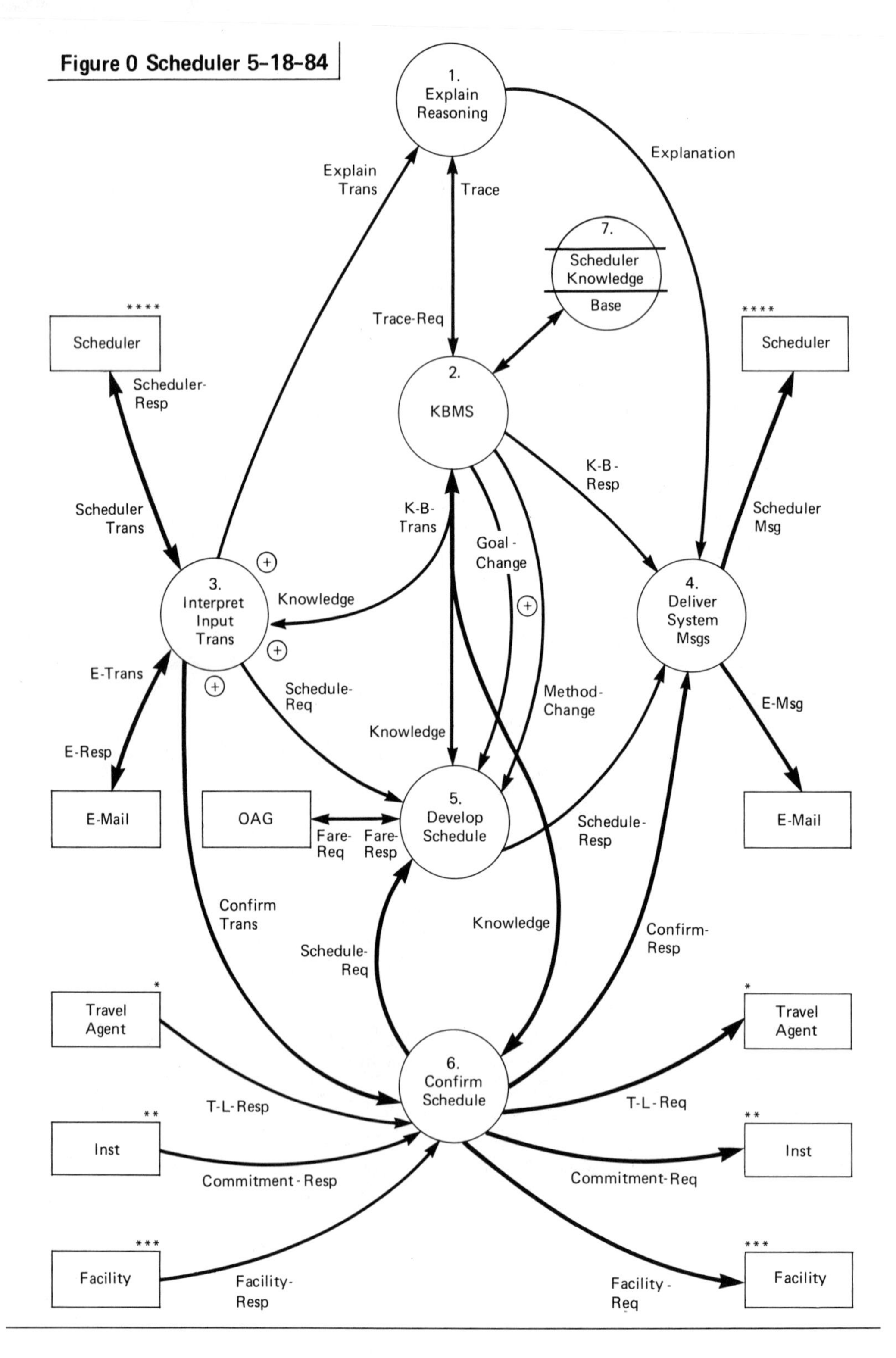

Figure 5-1d. Confirm-Trans Processing

sends a Schedule-Req to Develop-Schedule. Develop-Schedule generates a new set of alternative schedules and sends them to the user as it would in response to a Schedule-Req from the user.

5.2 E-Trans

The expert Scheduler is set-up so that anyone in the company, such as management or salespeople, can use the system to query the knowledge base and do some what-if scheduling. A salesperson might, for example, want to know whether it's possible to assign a particular instructor to a major account.

All of these other users access the Scheduler through the company's electronic mail (E-Mail) system. As we can see from the data dictionary, however, they are limited to making KB-Queries, using the system to develop tentative schedules, and entering Explain-Transactions; they cannot make any permanent changes to the knowledge base or commit to a schedule.

Except for these limitations, the processing of an E-Trans is identical to a Scheduler-Trans.

6. Develop-Schedule

We can now look at some details of the Scheduler to get a better idea of how knowledge-based processing is applied to this problem. First we'll consider the overall problem we're trying to solve with this process and some general thoughts about how it could be solved. Subsequently, we'll zoom-in on the details of Process 5 on DFD Figure 0, Develop-Schedule.

6.1 The Scheduling Problem

The general problem statement (see Chapter 4, Figures 4-1 and 4-2) says that we have a human resource of about 35 instructors to assign optimally to a very dynamic, running calendar of about 30 engagements per month. With this information we can make some interesting, though discouraging, calculations.

Specifically, if we have to evaluate every possible combination of instructors and engagements, then we are confronted with having to evaluate about 2.7×10^{37} possible arrangements every time we wish to create a schedule, and even though today's computers are very fast, it could easily take one microsecond to evaluate each combination. This means that developing one schedule would require about 8.4×10^{23} *years!* That's even slower than the traditional scheduling techniques of linear programming.

As a solution we might consider getting a machine that's a million times faster than today's machines, and one with a billion parallel processors working simultaneously on the problem. Such increases in horsepower are not out of the question in the next couple of decades. But even with this magnitude increase, one optimal schedule would still take over 800 million years to develop.

6.2 Solution by Constraint

At first blush, these numbers seem absurd. Yet it is these very orders of magnitude that for years prevented the automation of solutions to interesting expert problems, and that even now confront knowledge engineers in most expert system applications.

The first way to tackle such a size problem is to dramatically reduce the number of combinations we have to consider, an approach which every expert uses instinctively. In AI jargon, such limiting considerations are called **domain constraints,** and their judicious application has been the salvation of many expert systems. In our Scheduler, there are several domain constraints that can be applied.

The first arises from the realization that even though we do 30 engagements a month, no engagement extends over a weekend. This means that instead of trying to schedule a month's worth of engagements at a time, we can deal with the problem one week, or an average of about eight engagements, at a time. With this one domain constraint only, we have reduced the problem not by a factor of 4, but by a factor of 10^{26}. This means that on average we need to consider only about 9×10^{11} combinations which, with our one-microsecond computer, will require only 4.4 hours per schedule.

We might now observe that, in any week's worth of engagements, only about 15 of the instructors are well-suited to undertake them. This may result from instructors being unavailable or having the wrong background or training. This second constraint reduces the number of combinations to 2.6×10^8, which could be handled in about 4.3 minutes.

Additionally, in any given week, there will be probably at least two de facto assignments, which reduces the number of combinations to 1.2×10^6, or 1.2 seconds of processing time. Further, in each week one combination of instructor and engagement which is prohibited because of customer, salesperson, or instructor preferences. This latter consideration reduces the combinations to 8.2×10^5, or .82 seconds per optimal schedule. Such a scheduling time should be well within the limits of acceptability for the human scheduler's task requirements.

As a final attempt to reduce the problem, a way to evaluate the most likely combinations first would probably achieve another dramatic, though hard-to-estimate, reduction in the time required to produce one optimal schedule. Techniques for efficiently searching problem spaces have been applied by AI researchers and others for many years, particularly in game-playing systems. In this regard, we can observe that searching for optimal solutions is an underpinning also of the techniques of linear programming mentioned earlier.

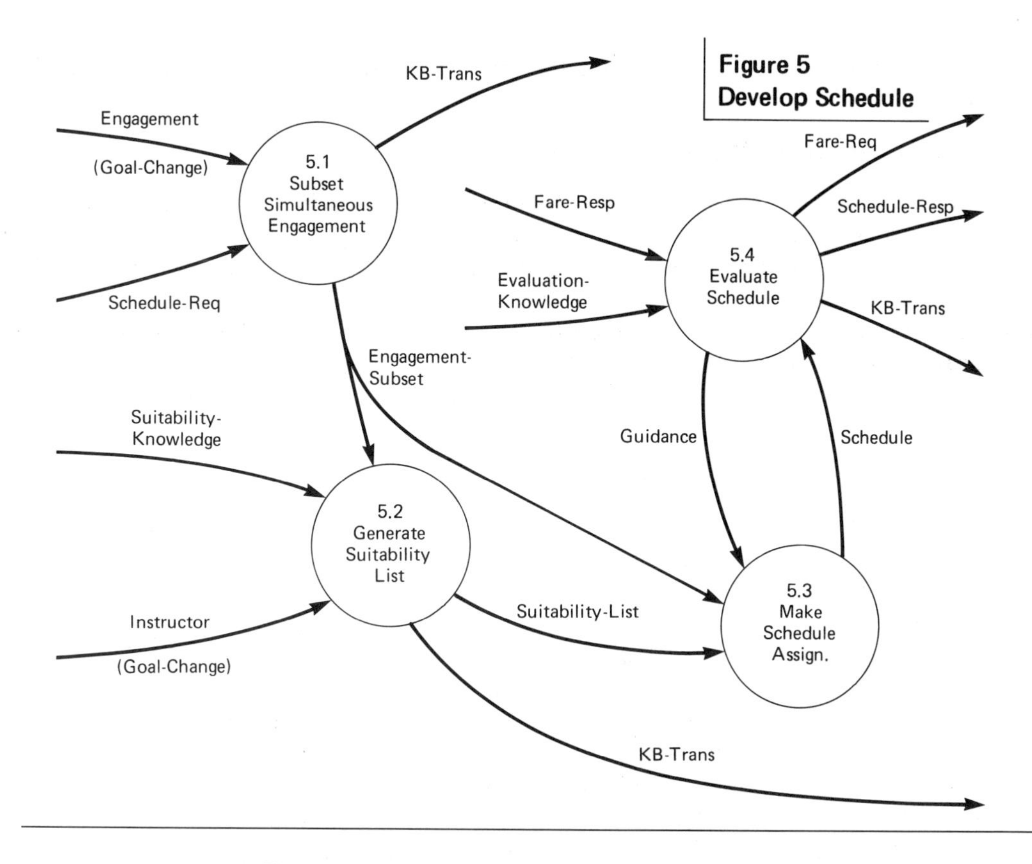

Figure 5-3. DFD Figure 5. Develop-Schedule Processing

6.3 Form Follows Function

Figure 5-3, DFD Figure 5, shows the details of the Develop-Schedule process and consists of four processes whose names are related to the domain constraints we have been discussing. This should come as no surprise, since our method of solving the scheduling problem suggested that we apply various domain constraints, and then search for an optimal solution by generating a schedule and evaluating it against some predetermined, and possibly heuristic, criteria.

6.4 Subset Simultaneous Engagements

There is nothing particularly advanced about the technology required for Process 5.1, Subset-Simultaneous-Engagements. Its job is simply to look at the dates of all pending engagements and group those which occur within a single calendar week. When such a grouping is established, it is sent both to Process 5.2, Generate-Suitability-List, and Process 5.3, Make-Schedule-Assignment.

For this discussion, let's say we have only four engagements in a particular week.

To further simplify the discussion, let's assume there are only four instructors to be scheduled.

6.5 Generate Suitability List

Process 5.2, Generate-Suitability-List is a bit more interesting since it appears to use some as-yet-unspecified knowledge from the KBMS.

The goal of this process is to create a list of all engagements and, for each engagement, provide a list of the instructors who are suited to teach it. We would like to include also some indication of how suitable each instructor is for that engagement—let's say a **suitability number** between 0 and 1. The questions are: How do we go about deciding whether or not an instructor is suitable, and how do we assign a useful suitability indicator?

6.5.1 Suitability Knowledge Structure

The method of Generate-Suitability-List is to follow the example of the human scheduler and make the suitability assignment based on heuristic rules about instructors and engagements, along with some indication of our certainty that each rule is correct. For example, one such rule might be: "If the customer is MMS, don't send old Foobong (1.0)." In this example, the (1.0) says that on a scale from 0 to 1.0, we are absolutely certain that we should not send Foobong to MMS.

In the same way that we developed a structure for KBE in Chapter 3, we must develop a structure for the knowledge associated with the inference processor, Generate-Suitability-List. This means discovering the factors and values which go into making suitability decisions. Rather than repeat the details of developing such a structure as was done for KBE, I simply refer you to Figure 5-4, Suitability Knowledge Structure, where the different decisions, factors, and values are listed. This is a fairly small set of factors, and experience with Scheduler may certainly change both the factors and their values.

The structure chosen for this example is fairly simple even though there are multiple decisions as there were in KBE and other factors may be involved in the decision. Instructor Background and Customer Background have been combined in a separate decision to produce a single factor called Background Match in this structure.

Also, multiple factors go into determining the Instructor's Attitude about doing an assignment, including time of year, city, and customer. Another approach to this would be simply to ask each instructor about each engagement. This seems to be an inefficient method compared with including some heuristics from the instructors about things they like and don't like.

In any system, it is desirable to have more than a paper model of the system's behavior. This is not always crucial when the system is making decisions based on well-determined conditions. However, where the system's performance is dependent on even the relatively few combinations of heuristic rules we have in Scheduler, it is usually very

```
DECISION:

Inst. Suitability    ASSIGN         DON'T ASSIGN

FACTORS               VALUES

Inst. Qualification  SRD-TRAINEE    SRD-INSTRUCTOR  SRD-SR.-INSTRUCTOR
                     PDQ-TRAINEE    PDQ-INSTRUCTOR  PDQ-SR.-INSTRUCTOR
                     (All other Inst. Levels) . . .
Inst. Availability   NO             YES
Inst. Attitude       REFUSES        WILLING         PREFERS
Sales Attitude       REFUSES        WILLING         PREFERS
Customer Attitude    REFUSES        WILLING         PREFERS
Background Match     POOR           FAIR            GOOD
Travel Expense   VERY LOW           LOW  MODERATE   HIGH   VERY HIGH

DECISION:

Inst. Attitude       REFUSES        WILLING         PREFERS

FACTORS               VALUES

Instructor           (All Instructors)
Customer             (All Customers)
Seminar              (All Seminars)
Seminar City         (All Cities)
Season               SPRING         SUMMER          AUTUMN   WINTER

DECISION:

Background Match     POOR           FAIR            GOOD

FACTORS               VALUES

Inst. Background      MGMT          ANALYSIS        PROGRAMMING
Audience Background   MGMT          ANALYSIS        PROGRAMMING
```

Figure 5-4. Suitability Knowledge Structure

difficult, if not impossible, for people to assess performance without seeing the system in action.

When choosing values for such factors as Instructor's Attitude and the Customer and Sales Preferences it is tempting to use a continuous scale from 0 to 1. This really allows for much too fine a specification of these very judgmental factors, and we are probably better off using the three discrete values shown. If finer resolution becomes necessary, we can either change the values, or we can include uncertainty in the statement of factors. For example, we could say that instructor REFUSES (50%) and is WILLING (50%). In any case, the choice made here to use three values is not very limiting.

The single decision, ASSIGN/DON'T ASSIGN, will in most cases be qualified by a suitability number which is less than 1.0.

6.5.2 The Suitability Number

From Figure 5-4 we see that the human scheduler uses several criteria in deciding what instructors will work with which engagements. These include fairly well-defined criteria such as the instructor's background and experience in teaching a particular seminar. There are also some very badly defined criteria, such as those which relate to the *feelings* of the customer and the salespeople about a particular instructor, not to mention instructors' *willingness* to go certain places.

One way of deriving a suitability number could be to combine the certainties found in the rules used to decide suitability. For example, the following two rules apply to a particular engagement-instructor evaluation:

```
IF THE INSTRUCTOR IS KELLER
THE SEASON IS SPRING
THE ENGAGEMENT CITY IS WASHINGTON, DC
THEN ASSIGN THE INSTRUCTOR (.85)

IF THE SKILLS (for this seminar) IS TRAINEE
THEN ASSIGN THE INSTRUCTOR (.3)
```

One way to combine the certainties of these two rules might be to simply average them, arriving at an overall suitability number of .58. Another method might be to use the smallest certainty as the certainty of the result, in this case .3. I shall defer a final decision on this for now, and in fact the best method for combining certainties may have to be changed as we gain experience in building and using the Scheduler.

6.5.3 The Suitability List

Regardless of how the suitability number is derived, the output of Generate-Suitability-List is a list of engagements and instructors as shown in Figure 5-5. The list shows the engagement number followed by a list of all the instructors who are suited to that engagement, with a number indicating how suitable they are. The process has also been thoughtful enough to list the instructors in descending order of suitability.

In the process of assigning suitabilities, a further simplification has occurred: Keller is the only instructor suitable for engagement #4. This becomes a de facto assignment and need not be considered further in generating possible schedules. The information about when an assignment is de factor is also heuristic, and might be contained in a part of the knowledge base which has rules about how rules are applied.

6.5.4 A Mini-Spec for Generate-Suitability-List

We now become aware once again that the DFD itself does not actually contain a specification of anything. Its only purpose is to partition the large Scheduler system into much smaller mini-systems and to show the information interfaces between them. It is in describing the workings of each process on the DFD that we are writing specifications, what I have been referring to as mini-specs.

As we begin to look at the details of processing in the Scheduler, we must observe once again the split personality of expert mini-systems. It may not be appropriate to discuss knowledge structure or training the system to make decisions in connection with Generate-Suitability-List. The reason is that *this process itself does not contain the expert knowledge—it is merely the inference engine which uses suitability heuristics stored in the knowledge base* shown on DFD Figure 0 as Scheduler-Knowledge-Base, Process 7.

```
#1 ((Zells .9) (Benson .85))

#2 ((Zells .98) (Benson .9) (Yourdon .7))

#3 ((Zells .85) (Yourdon .8))

#4 ((Keller .9))
```

Figure 5-5. Suitability List

From this point of view, Generate-Suitability-List is only part of the process of producing that list, the rest of the process being in the form of rules in the knowledge base. As you can see from Figure 5-6, one possible mini-spec for Generate-Suitability-List (discussed below), there is no mention of specific rules or data, only a statement of the way we want to use that information.

In this process and others, those rules become a dynamic part of the specification and eventually of the code which implements an intelligent processor. Process 5.2, Generate-Suitability-List is merely an inference engine which knows how to incorporate that knowledge into its processing to generate a suitability list. Its ability not only to use the knowledge to reach a decision, but also to produce a specific output closely related to the problem we're trying to solve suggests that it is a domain-specific inference engine, as opposed to a shell, which is independent of its domain of application.

Also as we shall see when we study DFD Figure 2, KBMS, even though the rules are used as part of the process of Generate-Suitability-List, those same rules are dealt with as data when they are added and modified by the KBMS.

6.5.5 So, What Goes Where?!

Unfortunately, the ambiguity of knowledge as data and knowledge as process can also cause some confusion about where to put which DFDs. After all, the rules are very closely related logically to the Generate-Suitability-List inference engine, and it might make sense to specify them along with the inference engine.

In fact, we will develop two specifications for the knowledge base component of Generate-Suitability-List—one for its data characteristics and one for its process characteristics—and these will be kept in different places. The logical knowledge structure and rules will be developed in connection with Process 5.2, the Generate-Suitability-List inference engine, while the data description will be drawn as a part of Process 7, Scheduler-Knowledge-Base.

This association was implied in Chapter 3 where we drew for the Determine-Worth process another level of DFD (Figure 3-7) showing both the local inference engine and the local knowledge base manager. In the Scheduler system things are not quite that clear-cut because we have chosen to use a global KBMS in spite of having several local inference engines.

Thus a better way to deal with Process 5.2 is to draw DFD Figure 5.2 (shown in text Figure 5-7) and use the mini-spec shown in Figure 5-6 as the mini-spec for Process 5.2.4. This DFD shows the central role played by the inference processor not only in managing the heuristics for any one mini-knowledge base, but also in managing the flow of data implicit in the DFD method. In actuality, managing the interplay of the three mini-knowledge bases in Figure 5-7 may require a more complex specification, or another level of DFD, for the inference engine.

As we continue to walk through DFD Figure 5 and others, it should become increasingly clear that Scheduler is not a single large knowledge-based application, but rather several smaller, independent, domain-specific inference processors using their own relatively independent, small knowledge bases. In each of those cases, we could draw diagrams similar to Figure 5-7.

```
For each combination of Engagement and Instructor
      Assign a Suitability-No using Suitability-Rules
If there is only one suitable instructor for an Engagement,
      Update the Schedule-Assignment
      Ignore that Engagement-Instructor-Pair
Output the Suitability-List
```

Figure 5-6. Generate-Suitability-List Mini-Spec

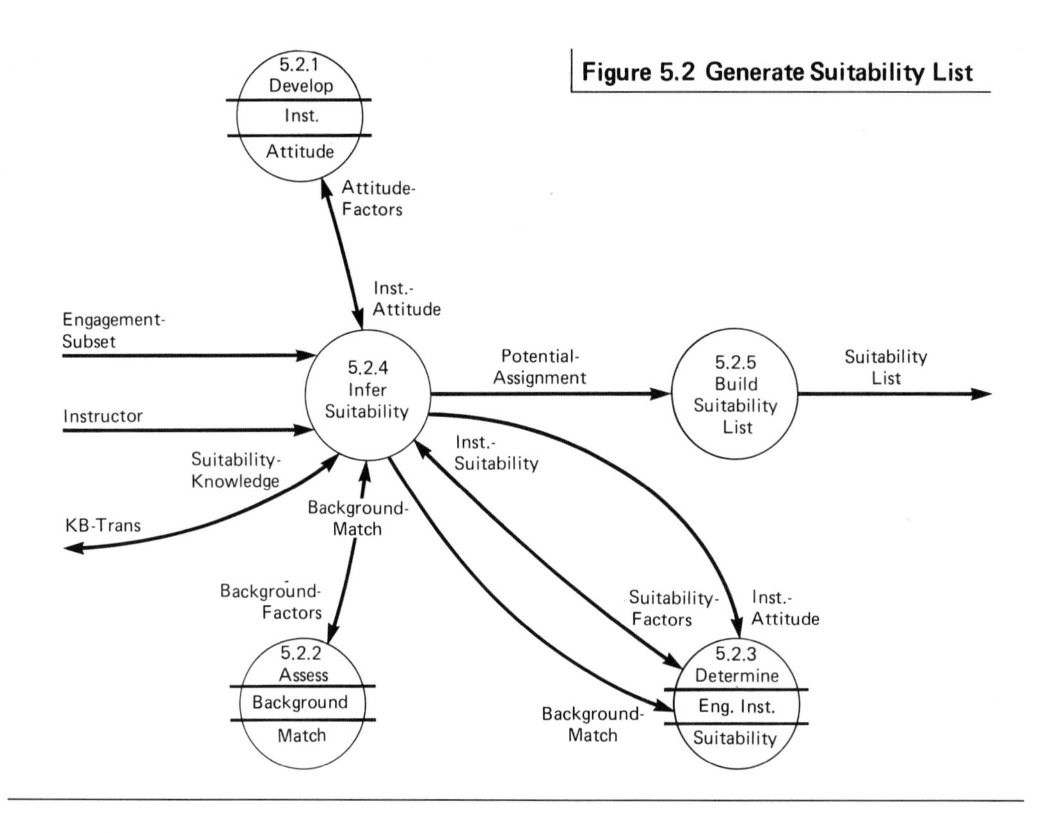

Figure 5-7. DFD Figure 5.2.

6.6 Finding the Optimal Schedule

6.6.1 The Schedule Tree

The suitability list can be viewed as a tree, as shown in Figure 5-8, with its root at the top of the figure. Below the root, each level in the tree represents one of the three engagements remaining to be scheduled after the defacto assignment of Keller to engagement #4 is removed.

At each level, the tree shows all possible combinations of instructors for that engagement based on choices made at the next higher level. Thus for engagement #1, tree level 1, the tree shows that either Zells or Benson can be chosen. At tree level 2, engagement #2, the choices are now limited depending on whether Zells or Benson is chosen for engagement #1. If Zells is chosen for #1, then Benson or Yourdon are available for #2; if Benson is chosen for #1, then Zells or Yourdon can be used for #2.

Each branch of the tree is numbered in arbitrary sequence as a convenience in referring to individual branches of the tree. In this representation, following a single set

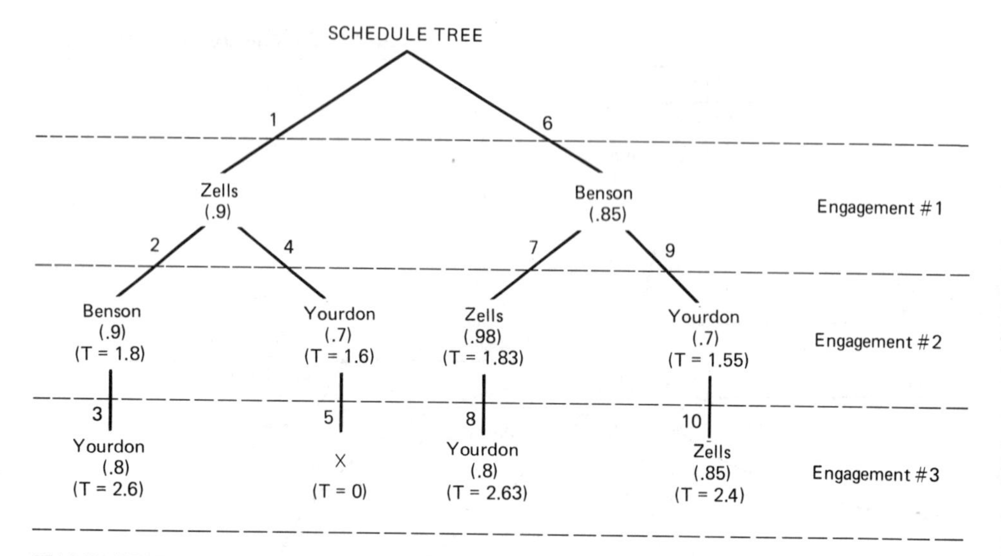

Figure 5-8.

of branches from the top to the bottom yields one possible schedule. One of the ways of doing this would be to take the right branch whenever there is a choice. For this tree this would mean following branches 6, 9, and 10 thus assigning Benson to engagement #1, Yourdon to #2, and Zells to #3.

6.6.2 Where's the AI?

Tree searching is used extensively in AI applications and has been used widely in mathematical optimization technologies for a long time. The AI here is not so much in generating a particular schedule from the suitability list, but in the heuristics for dynamically determining the best strategy to use at a given point in time based on the evaluation of schedules generated up to that point.

In addition to instructor names, the tree includes the suitability number for each assignment. Our goal is to find that route from the top to the bottom of the tree which will give us the largest number if we add up the suitability numbers at each level. The tree also shows at each node the total suitability accumulated to that point. Some other metric could be used, but this will do for now.

It is the job of Process 5.3, Make-Schedule-Assignment, which is a basic tree-maneuvering algorithm, and Process 5.4, Evaluate-Schedule, which is an inference engine, to work together and determine the best schedule or set of schedules. The strategy used to select a particular schedule, evaluate it, and provide guidance for choosing the next possible schedule is also heuristically driven.

The strategy could be as simple as evaluating all possible schedules and choosing the one with the highest number. It could also use a more complicated strategy which stops looking when a threshold suitability has been achieved. Guidance could include the possibility of using the same instructor for two engagements in the same week if they do not overlap and are in the same city.

7. The KBMS and the Knowledge Base

All during the discussion of Develop-Schedule we have referred to rules in the knowledge base and the use of those rules. Let's look now at some of the details of this basic KBMS and the knowledge it manages.

A KB-Trans is the data input to the KBMS. It can be generated by the user, by the Develop-Schedule process, or by the Confirm-Schedule process. Figure 5-9 (DFD Figure 2, KBMS), shows the four flavors of KB-Trans: Rule-Update, Check-Trans, KB-Query, and Frame-Update.

7.1 Rule Update

Process 2.3, Teach-Knowledge, is more than just an update processor for the rules which are a main part of the system's intelligence. It does include the functions of adding, deleting, and changing rules, but it also constellates Process 2.4, Check-Consistency, and Process 2.5, Self-Teach-Knowledge, which deal with additional implications of any change to the rules.

7.1.1 Where are the Rules?

Figure 5-10 (DFD Figure 7S, The Scheduler Knowledge Base) is actually not a DFD. It is what I call a Knowledge Base Structure Diagram and uses the knowledge base symbol (two horizontal lines through a circle) rather extensively. This symbol represents a knowledge base, some parts of which are static processing rules, some facts or data, and some dynamically both data and process.

This diagram represents different objects in the knowledge base and shows the logical access paths—*not* dataflows—between them. For example, the diagram shows that

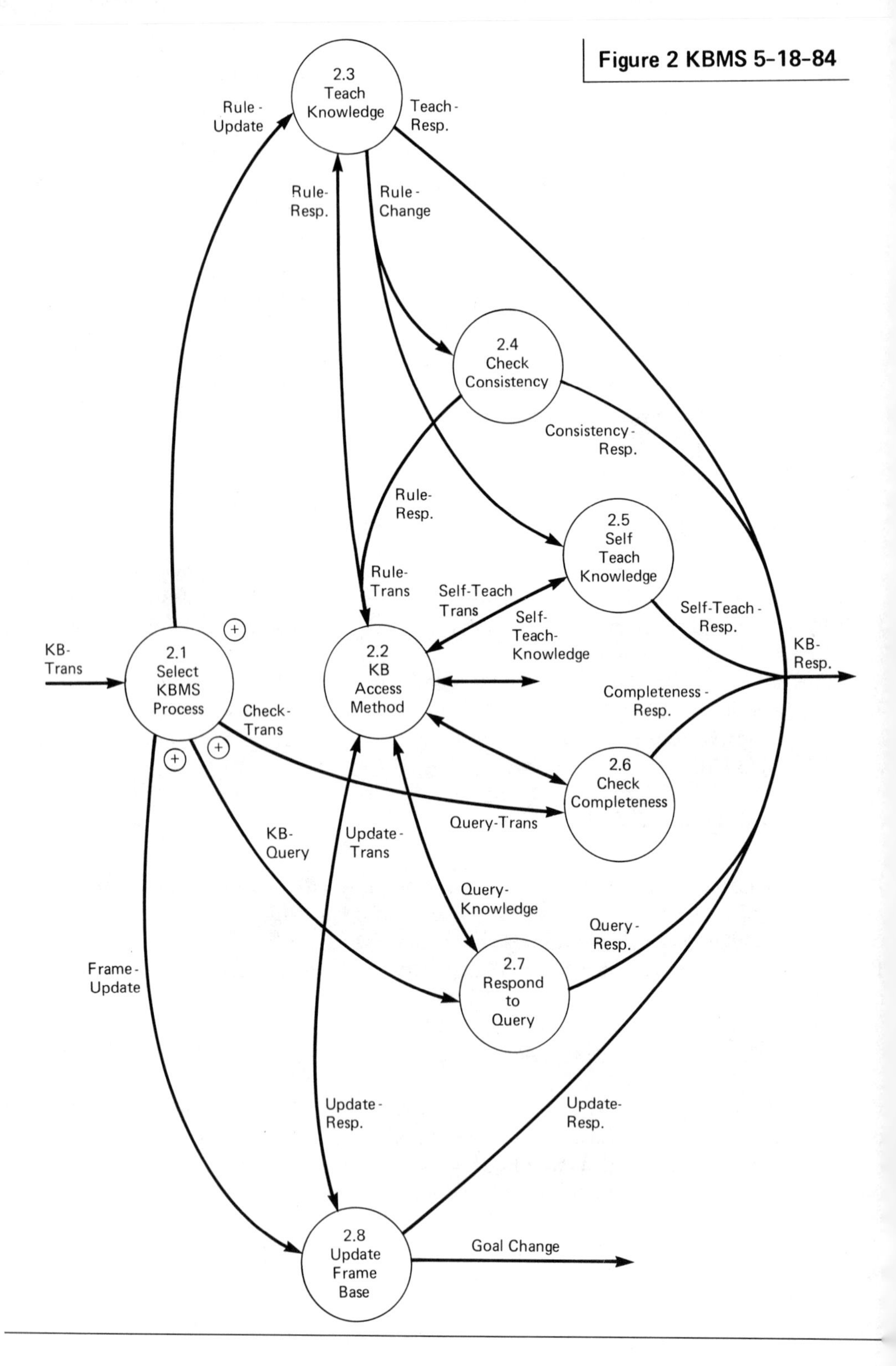

Figure 5-9. DFD Figure 2. KBMS Processing

if I know an Instructor (shown on the diagram as KB 7.1S) then I can access all of the rules pertaining to that instructor. Conversely, if I have an Instructor-Selection-Rule (KB 7.3S) then I can access all instructors referenced by that rule.

Further, the double arrow pointing in each direction between KB 7.1S and KB 7.3S suggests that for each instructor there may be many rules, and any rule may reference many instructors. A single head on the arrow, such as from KB 7.5S to KB 7.4S, implies

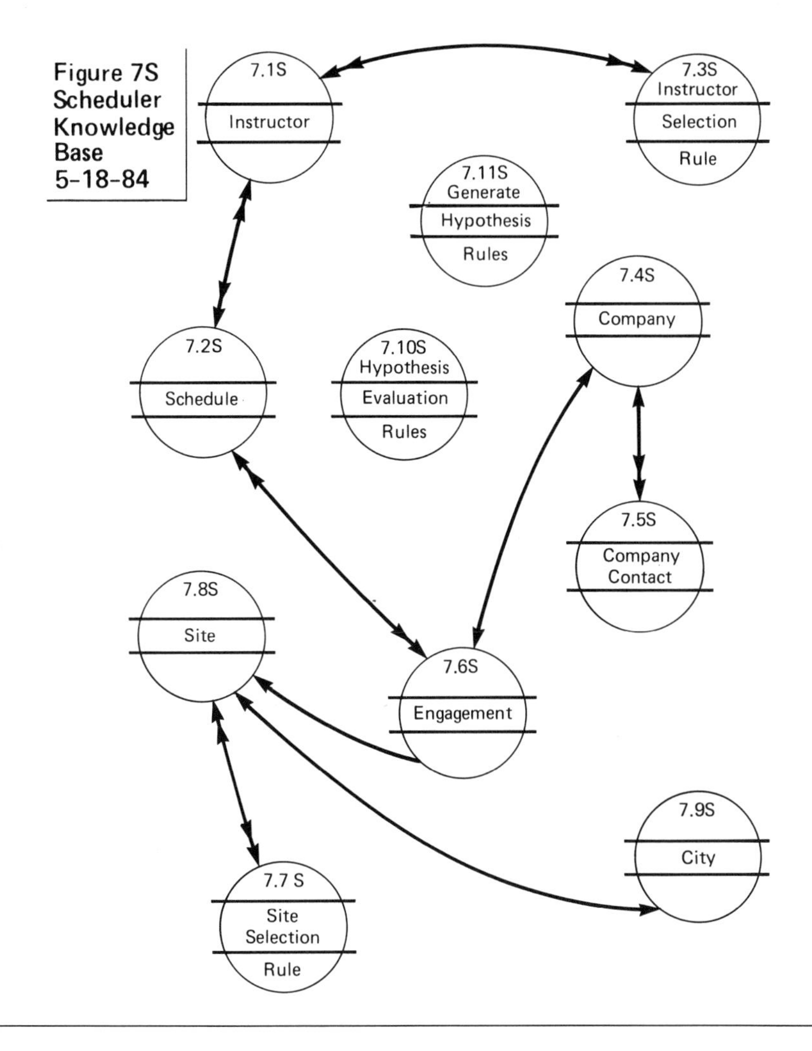

Figure 5-10. DFD Figure 7S. Knowledge Base Structure Diagram

a one-to-one relationship, in this case that for any Company-Contact there is only one Company.

This diagram is a complement to the DFD representation of this knowledge base in Figure 5-7. Once again, this dual representation results from the split personality which information frequently has in a knowledge-based system. That is, I want to treat a rule as data when I wish to modify it, and as process when I want to have an inference engine make a decision. The KB Structure Diagram (Figure 5-10) refers to knowledge as data, the DFD representation (Figure 5-7) to knowledge as process.

7.1.2 Consistency

Although the theoretical base for reasoning with rules lies in symbolic logic, which has a rigorous definition of consistency, **consistency checking** is informally used as a rubric for checking various logical problems which may occur. These include **logical consistency, contradiction, validity** and others.

For the Scheduler system, any change in the rule base may require checking for logical problems which are then communicated to the source of KB-Trans as a KB-Resp. Process 2.4, Check-Consistency, is responsible for this checking.

7.1.3 Learning

Process 2.5, Self-Teach-Knowledge, also receives notice of rule changes, although its job is somewhat different from Check-Consistency. Self-Teach-Knowledge means looking at the rule base as it is and trying to discover some general rules which will both encompass existing training and perhaps extend it.

For example, Scheduler might look at the rule base and find the following three rules:

```
  IF THE SEASON IS SPRING
  THE INSTRUCTOR IS JONES
     THE CITY IS TAMPA
THEN INST. ATTITUDE IS PREFERS

  IF THE SEASON IS SUMMER
  THE INSTRUCTOR IS JONES
     THE CITY IS TAMPA
THEN INST. ATTITUDE IS PREFERS

  IF THE SEASON IS AUTUMN
  THE INSTRUCTOR IS JONES
     THE CITY IS TAMPA
THEN INST. ATTITUDE IS PREFERS
```

As a first generalization, Self-Teach-Knowledge might do something very simple, such as replacing these three rules with a single rule which encompasses all of them:

```
IF THE SEASON IS SPRING THROUGH AUTUMN
   THE INSTRUCTOR IS JONES
   THE CITY IS TAMPA
THEN INST. ATTITUDE IS PREFERS
```

This type of self-teaching is generally called **rule compression** and can be useful in reducing the size of a knowledge base without altering its fundamental expertise.

A somewhat more interesting and creative approach to self-teaching would be to have the system realize that three of the four possible values of SEASON appear in rules in which the other conditions and the conclusion are the same; perhaps, it might conjecture, the same situation applies to the fourth value of SEASON, WINTER, as well. At this point, Self-Teach-Knowledge proposes this extended rule and waits for authority to add it to the rule base:

```
IF THE INSTRUCTOR IS JONES
   THE CITY IS TAMPA
THEN INST. ATTITUDE IS PREFERS
```

As suggested in earlier chapters, the subject of systems which learn by themselves from their experience has not advanced as quickly as other areas of machine intelligence. It remains something of a stumbling block to efficient knowledge acquisition. As we shall see in Section 8 of this chapter, Confirm-Schedule makes yet another attempt to learn from experience by evaluating the system's recommendations vis-à-vis the human scheduler's choices.

7.2 Check-Trans

Based on the lists of factors and values in the various decisions made by Scheduler, the system has a pretty good idea that knowing everything about the scheduling situation would mean having rules or training cases which give it an explicit decision for every possible combination of factor values.

The job of Process 2.6, Check-Completeness, is to compare the system's current rules and training cases to the maximum knowledge it could have in this domain and make an assessment of how evenly distributed its knowledge is. If there are areas that are very weak—that is, very dissimilar to previous training—Check-Completeness may suggest training cases for the user to enter.

7.3 Frame-Update

We have chosen to represent all stored information in our system using a flexible data structure call a **frame.** Each entity on Figure 5-10 (DFD Figure 7S) could be represented as a frame structure[8] which can be modified by adding or deleting data or by actually changing the structure. Process 2.8, Update-Frame-Base, is responsible for doing this general kind of change.

The heuristic rules used by Generate-Suitability-List, for example, are actually part of the frame-base, although we have specified separate processes for adding, changing, and deleting the rule frames. This is largely because of the need to check their consistency and use them in self-learning.

A frame-base is like a relational database in many ways, although much more flexible in the kinds of information it can store. In particular, each type of frame, roughly equivalent to a database logical record type, contains not only fields, called **slots,** of static data, but may also contain default values for slots or even procedures, called **daemons** or **active values,** which are activated when certain conditions in the frame are true.

It is through the Frame-Update transaction that all information, other than rules, is maintained in the Scheduler knowledge base. This includes such things as instructor availability, seminar site descriptions, and even the current dynamic schedule itself. When a change is made to the frame knowledge base—adding an engagement for example—this information is sent to Develop-Schedule as a Goal-Change and the scheduling process is invoked to deal with the modified information.

7.4 KB-Query

A generalized ad hoc query facility is available through Process 2.7, Respond-to-Query, for any of the other processes which require information, rules, or data from the Scheduler Knowledge Base. By the time the KB-Query type of KB-Trans arrives at Respond-to-Query it has been transformed from what may have originally been a natural English query into a completely formal syntax.

7.5 Accessing the Knowledge Base

All of KBMS functions access the physical files of information through Process 2.2, KB-Access-Method. As with any database manager, there is a set of functions whose job is to perform the physical reading and writing of information in response to a relatively small set of basic commands.

The KB-Access-Method is such a set of functions, except that here they must deal with the nuances of information stored in frames.[9] This means providing a control structure in which the daemons mentioned earlier are activated when appropriate and other functions relating different types of frames are performed.

[8] Refer to Chapter 8 for a more complete discussion of representing knowledge in frames.

[9] A basic set of frame access functions is discussed in Chapter 8.

8. Confirm Schedule - DFD Figure 6

Once the user has decided upon a satisfactory schedule alternative, a Confirm-Trans is sent to DFD Figure 0's Process 6, Confirm-Schedule. The details of Confirm-Schedule are shown as Figure 5-11 (DFD Figure 6). As you can see, the Confirm-Trans actually goes into each of the four different processes shown on this diagram.

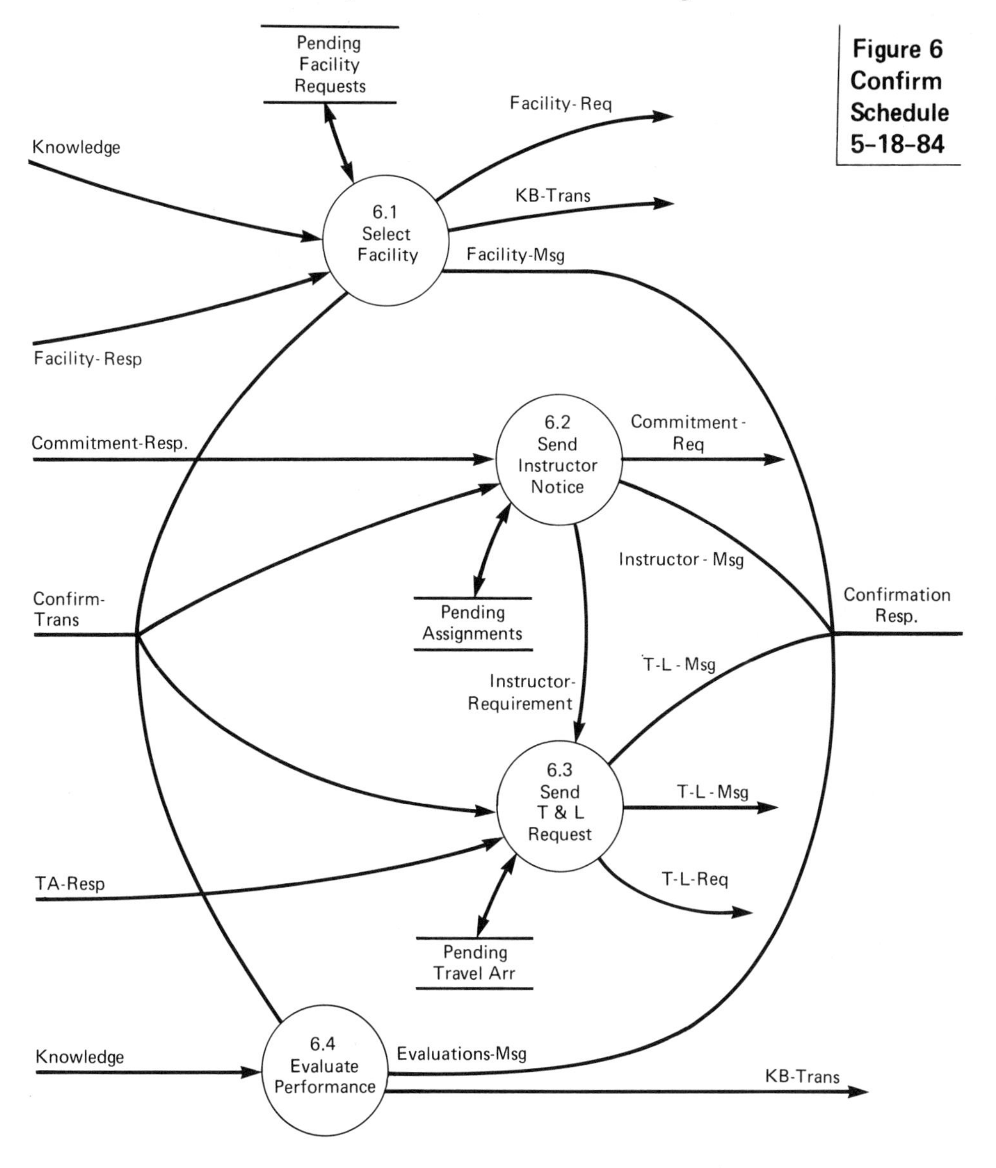

Figure 5-11. DFD Figure 6. Confirm Schedule Processing

8.1 Housekeeping Functions

Process 6.2, Send-Instructor-Notice, and Process 6.3, Send-T&L-Request, are little more than housekeeping functions. They have been told who is going where and when, and their task is to confirm those arrangements with the Instructor and a travel agent. These functions also must remain aware of pending arrangements and make the proper notifications if arrangements fail. Basically these are very straightforward processes.

8.2 Select Facility - Another Inference Engine

One might think that Process 6.1, Select-Facility, is also merely a housekeeping function. However, in any engagement locale there may be many facilities, each of which has certain size and service capabilities. In addition, for facilities at which past engagements have been held, instructors' comments may be used heuristically in deciding whether or not to use the facility again. Also, we may wish to make a deal with a customer in which a certain number of seats in a seminar are exchanged for use of the customer's classroom facilities.

So Select-Facility contains yet another inference engine which must access rules from the Scheduler Knowledge Base in evaluating facilities. Each of the inference engines encountered so far—such as Generate-Suitability-List or Select-Facility—is an instance of a domain-specific expert system because, although it uses external rules to make decisions, the actions it takes and the way each uses its inputs is specific to its domain.

8.3 Evaluate Performance

Process 6.4, Evaluate-Performance, is a cousin of the Self-Teach-Knowledge process in that its goal is to improve the expert performance of the Scheduler system. This is another instance where the Scheduler attempts to learn from experience. It does this by comparing its own first-choice schedule with the schedule chosen by the user. If they are the same, then everything is OK and Evaluate-Performance believes the system is performing up to standards.

However, if the user chooses another alternative, this is cause for self-study and perhaps revision of some parts of the knowledge base. The system's first action is probably to ask the user why she chose another alternative, and perhaps ask for some additional rules to be added to the Evaluate-Schedule part of the KB. Evaluate-Performance might go even further by formulating new rules based on the differences between the two schedules and proposing these new rules to the user as possible additions to the KB.

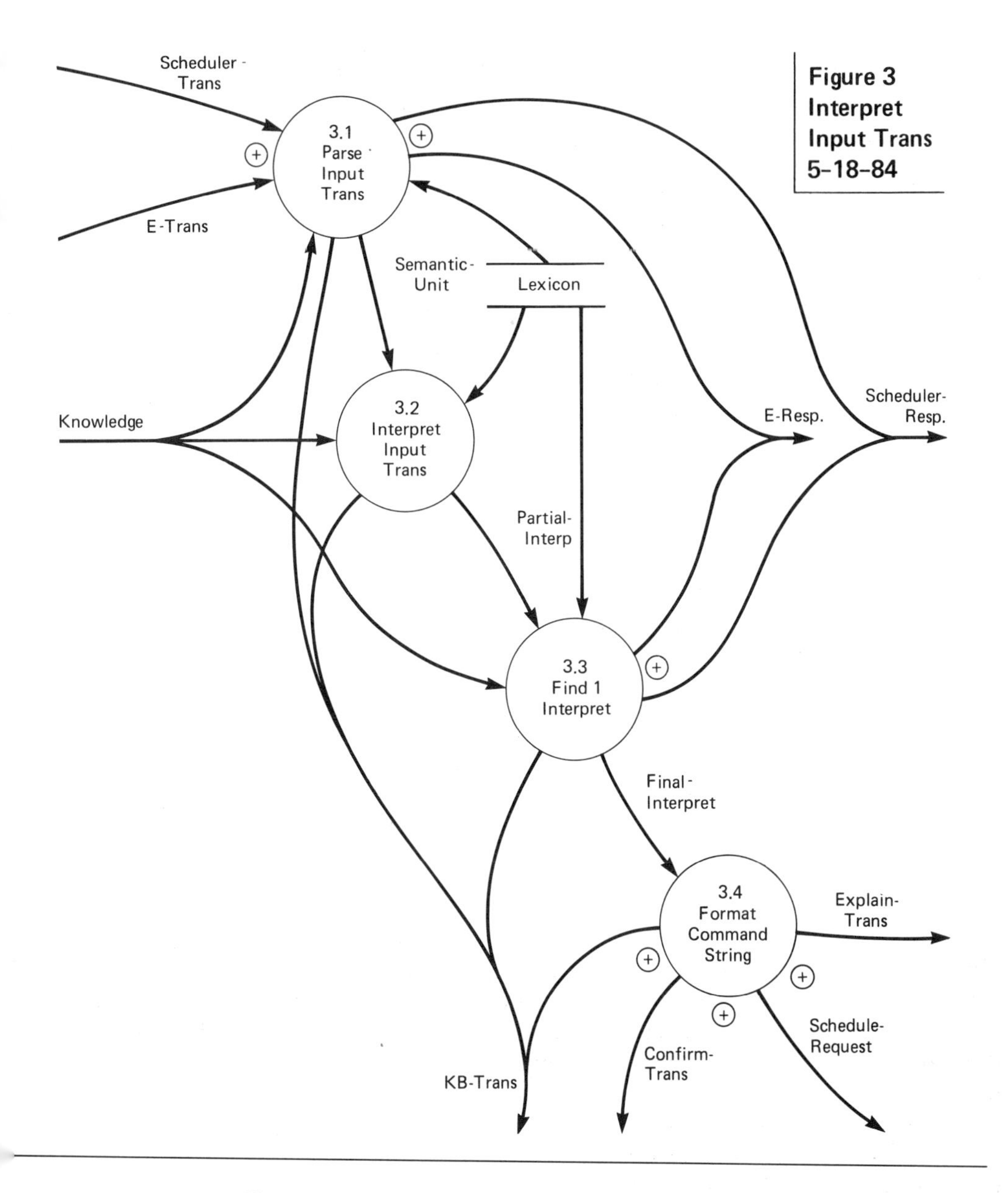

Figure 5-12. DFD Figure 3. Natural English Processing

9. Natural English Input

The interpretation of natural English in a limited domain such as scheduling is discussed in Chapter 9. For our purposes we see Figure 5-12 (DFD Figure 3, Interpret-Input-Trans) as a special instance of a natural English front-end system.

The interpretation methodology indicated by Processes 3.1, 3.2, and 3.3 is sophisticated enough to deal with issues of ambiguity, word order, and pronoun references, it is a well-known technology. The intent of every interpretation is to format a command string for one of the system's functional transactions, shown in Figure 5-12 as outputs from Process 3.4.

10. Completing Analysis

This completes our review of the issues surrounding the use of knowledge-based processing in the specification of Scheduler, which in the real world must be considerably more complete both in terms of functions and definitions in the data dictionary. Nonetheless, from the specification presented it should be clear how AI technology blends with more traditional approaches and how the DFD technique helps identify those areas where AI is required.

However, the job is not complete. At the beginning of the discussion we observed that this specification represents the New Logical phase of the system. This means that, even though most of the system's functions were described as though they were automated, we have specified functions, but not how they are to be implemented. We must now deal with the extent to which the system will be automated.

10.1 The Automation Boundary

Every function on the DFDs could be done manually (as in the current system), through some automated processor, or by a combination, which is more likely. The last step of analysis is to determine how much of which functions are going to be automated. This automation will be shown by adding to the New Logical DFDs processors to do low-level physical things like transporting information—data or knowledge—across boundaries: physical, political, geographical, or any others you can think of. The New Physical DFDs will specify all those physical interfaces, without which we have no system. In the real world this is no trivial task, since different degrees of automation will affect the cost and duration of the project as well as the amount of work the human expert will still have to do.

For example, referring to Figure 5-3, Develop-Schedule, we might decide the Process 5.1, Subset-Simultaneous-Engagements, will be done manually. Perhaps even Process 5.4, Evaluate-Schedule, could be done by the human scheduler.

Even at this New Physical stage, however, we may find a need for some knowledge-based processing. Let's say, for example, that Evaluate-Schedule, Process 5.4, were automated so that a human did the actual evaluation, but accessing the Official Airline Guide's Electronic Edition was automated.

We might need an expert system in this situation to automatically dial-up and establish a viable connection with the OAG service, handling in an expert way all the strange things that can happen on a telephone line.

11. Conclusion

All during analysis we have been stating policies about the way seminar scheduling should be done and provided some knowledge, in the form of rules, to specify the decisions.

However, nowhere have we talked about how these policies will be implemented. Traditionally, the design phase addresses the means and methods used to make a system a reality. Rather than treating design issues in detail, however, we'll look at one of the more exciting and popular aspects of modern system development: prototyping. The many new tools for prototyping have changed the design situation somewhat. The next chapter looks more closely at the issues surrounding the practical use of prototyping in knowledge-based system development and presents one way of doing it that is closely related to DFDs.

Chapter 6
Structured Analysis & PROLOG
Realistic Prototyping

In this chapter we shall discuss the uses of prototyping and the relatively new programming language called PROLOG. PROLOG is the first seriously popular, powerful language that embodies an approach which is quite different from COBOL, PL/I, or even LISP: logic programming.[1]

PROLOG and its yet-to-be-invented derivative languages are likely to have an important impact on the traditional usage of structured analysis and on the design and construction phases of a system project. It introduces a new way to think about system development without discarding the best features of our tradition. We shall discuss those features which make PROLOG a uniquely natural way of implementing at least prototypes, if not production versions, of the knowledge-based mini-systems we have identified during structured analysis.

In the following discussion, I make some fundamental, though in some cases perhaps unwarranted, assumptions:

• First, I assume that your goal from the beginning has been to apply knowledge-based processing in relatively simple, though interesting and useful, ways at first and to learn a lot in the process.

• Second, I assume that you will not try to do the knowledge-based parts of your system in COBOL. It's possible to write knowledge-based applications in COBOL, but

[1] The idea of logic programming has been around since the early 1970s at least, but had not caught on until PROLOG came on the scene a few years ago.

it's not easy, it's not fun, and it will detract from your learning experience. Furthermore, you probably won't be able to find any systems people who are interested in doing it. At least try to work with FORTRAN or BASIC, preferably PASCAL or C, and best of all LISP or PROLOG.

- Third, I assume that you do not have available a sophisticated knowledge engineering environment such as ART[2] or KEE.[3] If you do, so much the better. In my opinion, having such tools does not obviate the need for a thorough analysis, but it can make analysis, prototyping, design, and construction much easier for any system, not just knowledge-based systems.
- Fourth, I assume that your company has a traditional system environment such as IBM mainframes or PCs.
- Fifth, I assume that you have the options both of acquiring a reasonable amount of new software and of giving a few individuals a limited amount of training in AI techniques.

1. Working Models—the Rapid Prototype

1.1 What is a Prototype?

Without some qualification, the word "prototype" has a very elastic meaning. It can be for example, a "paper prototype" which models, perhaps very accurately, the way a system would work if we were to implement it. It might be also the first approximately working model that we have of some system, the "early prototype," or the somewhat improved "working prototype." It can also be used for that "advanced prototype" version which, with just a few more bells and whistles, will be ready for production use.

In all cases, however, it seems to carry a connotation of being a model, of being somewhat unfinished, of somehow falling short of the real thing. So in prototyping a computer system, we may make some simplifying assumptions about data structures, relationships, or processing. The prototype is not the finished product, but it can be a very useful approximation to that final goal.

1.2 The Rapid Prototype

"Rapid prototyping" is popular jargon these days, and it refers to a computer simulation of a system which typically lies somewhere between the "early" and "working" prototype stages. There are many good reasons for wanting to create rapid prototypes.

The useful DFD modeling techniques of structured analysis even today and are usually paper models and require frequent redrawing by hand, a cumbersome process to say the least. The new generation of knowledge engineering tools (discussed below)

[2] The Advanced Reasoning Tool is developed and marketed by Inference Corp. of Los Angeles, CA
[3] The Knowledge Engineering Environment (KEE) is developed and marketed by IntelliCorp of Menlo Park, CA

brings with it the motivation and the means to extend paper modeling ideas and produce incrementally evolved working models of the system as part of the specification process—in other words, a **rapid prototype.**

One of the appeals of rapid prototyping is that it acknowledges the human's intrinsically poor performance at doing things right the first time, and our historically proven gift for taking something imperfect and refining it to a highly polished state over time. Moreover, systems people want to start making something work early in a project, rather than waiting until the end of lengthy paper analysis and design phases to begin coding. One of the principles made clear by AI research over the last 30 years is that the only way to prove the efficacy of a computational theory, which includes a system specification, is to write a program to make it work. It may be because of the need for immediate proof that the rapid prototype concept has evolved out of the AI discipline.

For whatever reason it is here, this approach offers a new opportunity for data processing to adapt system development methodologies to include some capabilities for rapid prototyping. Rapid prototyping has the potential to produce more complex systems, more quickly and successfully than ever before.

1.3 The Dangerous Prototype

The typical non-data processing end-user is easily fooled by the way things appear to be on a computer screen. When a computer appears to do what a user wants, it is natural for the user to assume that the computer is actually doing it in a well-planned, efficient manner.

Consequently, a good prototype can easily be mistaken for the real thing. In fact, more than one end-user has refused to fund development of a production version of a system whose prototype was a bit too impressive.

Thus the advent of new prototyping tools holds the promise of being able to get the correct idea of a system early, but it also can be an excuse not to do the careful analysis and planning required in any system. Prototyping can be a wonderful enhancement to structured analysis, but *not* a replacement for it. Even the most sophisticated modeling tools cannot replace human beings interacting to decide what needs to be done and how to do it.

2. Modeling Tools

Many approaches are emerging to give systems people some automated help in specifying and implementing new systems. The ones discussed here are generally intended to give either clerical assistance or assistance in creating working models. Ideally we would like a tool which does both and can subsequently easily extend prototypical systems to their production equivalents.

2.1 Development Environments

Whatever modeling process we choose, it will exist in some development environment: the sum total of hardware and software that systems people have to do their job. A lot of interesting work can be done in relatively simple environments, but the new generation of software and hardware prompts us to look at where we have come from and what will be possible and necessary in the future.

As COBOL programmers sit at their IBM 3270 terminals, they typically use a text editor to look through the source code and correct errors in the latest compile of the system they're building. When the changes have been made, they exit from the editor program and invoke the COBOL compiler and linkage editor. They then wait—perhaps seconds, but more likely minutes or hours—for the compiler job to be scheduled and run by the computer. If the program compiles and links without error, the programmer will run a JCL job stream to test the program, again perhaps waiting some time for the job to be scheduled and run. The cycle generally repeats many times before the code works as specified.

During this process, the programmer must be aware of every detail of the program's operation: array sizes, variable definitions, loop variables, and many others. This scenario describes the somewhat burdensome **development environment** in which most commercial programmers work these days.

AI programmers using a LISP machine have quite a different experience. In a LISP environment there may be no compiler, and the program editor is an integral part of the same system that runs the code. The program is written, entered, and debugged as a single integrated series of actions. When the program works, it will be submitted to a compiler once and then be used in a run-time production environment.

In addition, there may be an intelligent **programmer's apprentice**[4] facility which takes care of much of the bookkeeping that would be required in a COBOL program. Clerical issues related to array sizes, variable types, and the like are history to the AI development environment.

The environment may also contain debugged prototypes of frequently used types of program so that the programmer need only make changes peculiar to the system being developed. In addition, while testing the program, the system keeps a history of user inputs and their effects, which can be modified or repeated, and can even back out a series of inputs. A long list of "programmer-friendly" features is transforming the previously boring and frustrating task of producing a working program into a simple step in the creation of a large system.

This enlightened approach to the program development environment is generally quite expensive today, although the prices of AI hardware and software are dropping rapidly. Our purpose in discussing the following modeling tools is to put some different

[4] Most AI hardware manufacturers either already provide or plan to provide some such help for the programmer. The description is based on the Programmer's Assistant facility provided by the XEROX INTERLISP-D environment, which runs on their 1100 series AI workstations.

approaches in perspective and try to find sensible ways for any company to participate in the revolution of the Fifth Generation.

2.2 Fourth-Generation Languages

Most data processing shops associate rapid prototyping with the use of fourth-generation languages (4GLs) such as FOCUS, RAMIS, EASYTRIEVE, and countless others. In this context we are typically concerned with getting answers to ad hoc requests for information in a matter of a few hours or perhaps a few days.

The 4GL approach is a pathetically primitive approximation to prototyping because the 4GLs have very weak representation and system modeling capabilities. Thus they are suitable for a limited class of problems in which simple, well-determined processes are applied to well-defined data bases to produce well-defined reports. In addition, they tend to be quite cumbersome to use, make highly restrictive assumptions about the way information will be stored, and have no capability for extending prototypical systems into efficient production systems.

The main advantages of 4GLs are that most companies have at least one already and, for the limited class of problems they can handle, reasonable prototypes can be produced relatively easily.

2.3 The Analyst's Workbench

A more useful and interesting facet of the system modeling problem is being addressed by a type of system which might be called generically an analyst's workbench[5] which is heavily oriented toward providing automated clerical assistance in developing and maintaining DFD specifications. With their present capabilities, they can potentially break the paperwork bottleneck which has severely restricted the spread and effective use of structured analysis and design for more than a decade.

Unfortunately, these systems stop short of giving us what we really want, which is a DFD in which the primitive functions can actually be working models of the real thing. There is no inherent reason why working prototypical functional primitives could not be built into analyst's workbench systems and thus implement the rapid prototype concept for structured analysis directly; their early mission, however, seems to be the more clerical approach.

We can only hope that second-and third-generation versions of analyst's workbench systems allow more than paper prototypes.

[5] One such system, called *The Analyst Toolkit* is being marketed by Yourdon, Inc. of New York, N.Y. Another such system, *Excelerator,* is developed and marketed by Index Technology of Cambridge, MA. Many similar systems are expected to be available from other vendors in 1986-90.

2.4 Knowledge Modeling Systems

At the other end of the spectrum from 4GLs and the analyst's workbench are very sophisticated, and expensive, modeling systems such as KEE,[6] which are oriented toward

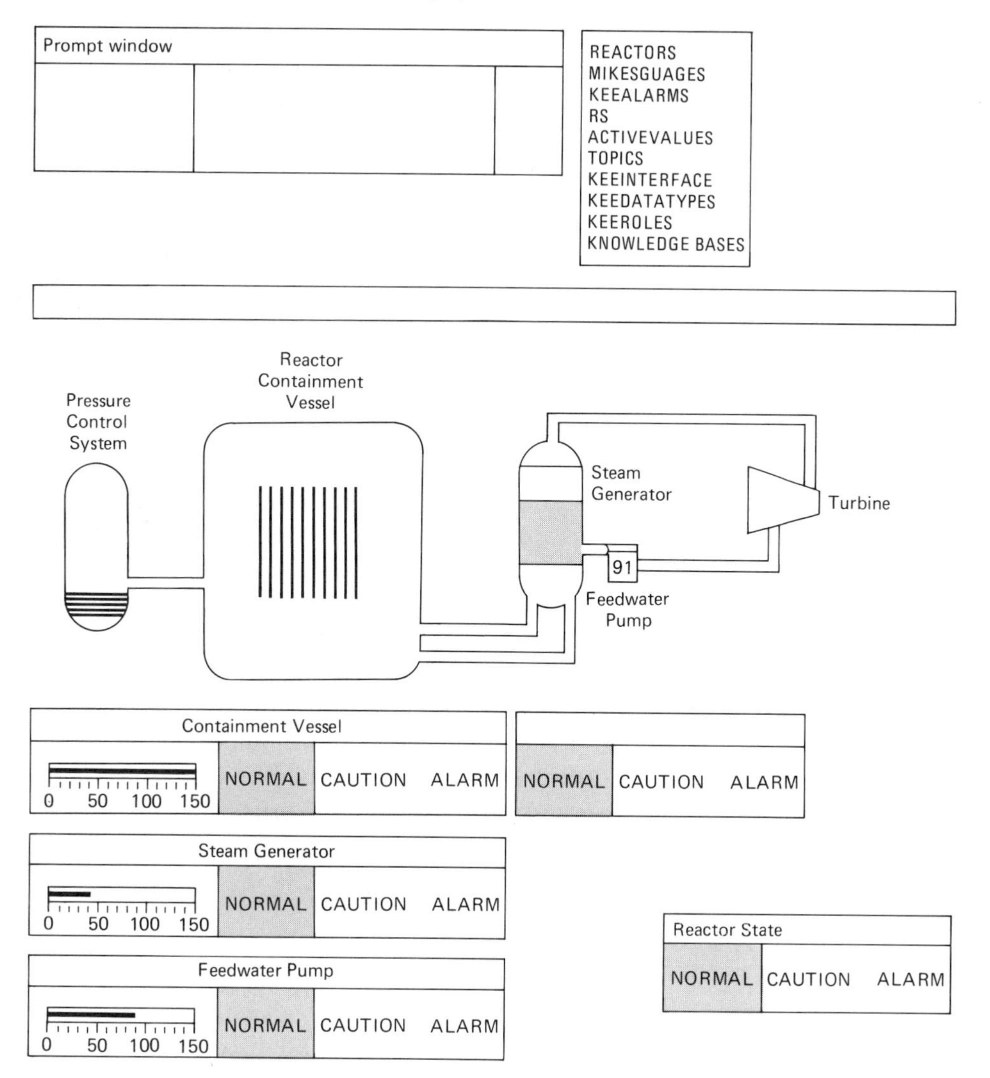

Figure 6.1. A control panel for the reactor system. Each gauge both reports the parameter value and can be modified by the user with a mouse. (Reprinted by permission of Thomas P. Kehler, John Kunz, and Michael D. Williams.)

[6] The Knowledge Engineering Environment (KEE) is developed and marketed by IntelliCorp of Menlo Park, Calif.

being an evolving system, possibly never reaching a static "production" state. This attitude is appropriate for an expert system development environment, since both human and machine expertise continue to evolve over a long period of time.

The KEE system itself consists of an effective combination of knowledge representation and reasoning techniques with an extensive capability to use very-high-resolution graphics.[7] This allows the user/analyst team to view instantly the effects of data or process changes made anywhere in the system as shown by Figure 6-1, which shows a control panel screen for a reactor system developed using KEE. As the user changes values in this prototype, the qualitative and quantitative effects are seen immediately on the screen.

Though it may not be apparent from this KEE system graphic, it is a highly data-driven representation of the system, as is the DFD. I mention this mainly to underscore the fact that structured analysis as it is now practiced by many companies can be carried easily into the next generation of systems. The difference is more in the implementation of modeling techniques than in the basic modeling concepts themselves.

2.5 A Basic Prototype

In doing the KBE system in Chapter 3, we actually used a prototyping method which combines the DFD paper model and an expert system shell[8] to prototype the various intelligent processes on the DFD. This approach, while not nearly as elaborate as the KEE prototype, has the advantages of being available on standard hardware at modest prices, and of providing early feedback to user and analyst not only on the knowledge base per se, but also on the inference techniques used by the shell and its scheme for handling uncertainty.

As you can see, there is quite a variety of choices when we talk about modeling or prototyping; some are very simple, others quite elaborate. An important element, common to all the techniques, is reinforcement of a lesson learned from structured analysis: system development is an iterative and highly interactive process of discovering the optimal blend of functional requirements and technological capabilities.

3. Programming Languages

Before we can fully appreciate the uniqueness of PROLOG, it's important to look at the different types of programming languages available today and discover those features of PROLOG which make it truly unique.

[7] Although there is nothing inherently "AI" about very-high-resolution graphics with extensive windowing capability, such features seem to be a hallmark of the new generation of AI workstation machines.

[8] In this case the TIMM system from General Research Corp. was used. Any of the countless other commercial shells could have fulfilled this simple prototyping function similarly.

3.1 What vs. How

The DFDs of the New Physical model which appear in the structured specification contain a large number of mini-systems, each of which has a mini-spec written for it. These mini-specs, however, seem to tell us only what the function does to convert its inputs into its outputs, not how to write a procedural program which will actually do the transformation.

For example, if a mini-system required sorting some data, the mini-spec would say something like: "Sort the Invoices file by Customer-Number and Date-Sent." It would probably not tell us which of many sorting algorithms should be used, and certainly not the actual procedure of that algorithm.

Similarly, for the Scheduler system, the mini-spec for Process 5.2.4, Infer-Suitability tells us *what* needs to be done to generate a Suitability-List, not necessarily *how* to do it. For example, one of the steps in its specification might say: "Backward chain through the Instructor-Selection-Rules." Or it could make an even higher-level statement such as: "For each Instructor-Engagement combination, evaluate the Instructor-Selection-Rules." In any case, it would almost certainly not tell us how backward chaining is done for the Generate-Suitability-List process.

In traditional structured systems, the design document includes both the architecture of the required program modules and module specifications, which go into enough algorithmic detail for a programmer to write a COBOL program to accomplish the required transformation. In doing these structured design documents we are required to translate the "what-needs-to-be-done" requirements of structured analysis into the "how-shall-we-do-it" needs of current programming languages and operating systems.

In many ways, it seems to be a waste of time to have to change the structured specification in any way. After all, it specifies everything we want to have done, it is the closest representation of the user's business processing we'll find, and we already have it. When it comes to maintaining the system it would be wonderful to have to change only one set of documents which would specify both the users' needs and the working system.

3.2 Description vs. Procedure

Clearly what is needed is a programming language which fairly directly implements equivalents of the structured English mini-specifications used during structured analysis. And that is exactly what PROLOG does.

3.2.1 Procedural Languages

No matter what language we use, whether it's COBOL, or LISP, or even natural English, programming means making the computer automatically execute some sequence of steps in achieving some solution. With all traditional programming languages, a program is a detailed list of steps to follow, i.e., a **procedure,** such as in the following BASIC program:

```
1 INPUT X,Y
2 IF X>Y THEN 5
3 PRINT Y
4 GOTO 6
5 PRINT X
6 END
```

This says that the first thing we want the machine to do is to read two numbers from the terminal and put them in places in memory called X and Y. Then we test to see if the first number is bigger than the second and, if so, take our next instruction from step #5, and so on. Procedural languages sound like a treasure hunt where we're always discovering some new clue in our path.

The problem with this level of programming detail is that the programmer has to keep track of many details of the machine's operation—data types, array sizes, and many others. As a result the programmer is easily distracted from the essence of the problem to be solved, and for very complex AI problems for example, may find the overhead of a language like COBOL to be an insurmountable obstacle.

3.2.2. Early Descriptive Languages

Some languages are less oriented toward procedure and more toward a **description** or declaration of the problem to be solved. That is, instead of telling how inputs are read, massaged, and printed out, we simply describe the logical relationship between inputs and the outputs into which they are transformed. One of the earliest attempts at descriptive programming was RPG, the Report Program Generator. With RPG, instead of writing assembly language programs, one simply described the layout of fields on a report, indicated where sub-totals were to be taken, what causes a page break, etc.

RPG has now given way to dozens of fourth-generation languages (4GLs) which are, at base, descriptive programming languages. For example, with 4GLs a user "programs" the computer not by stating procedures but rather by describing, for example, the fields we want to see printed on a report and their positions. The 4GL therefore, more closely approximates specifying *what* needs to be done rather than *how* it should be carried out.

However, the 4GLs are a fairly primitive attempt at descriptive programming. We can see from the example below that they include a rather cumbersome procedural component which the user needs to get reports that are more intricate than just listings of fields. In other words the 4GL's main goal is to describe relationships between inputs and outputs, but the scope of relationships it can describe is very small and must frequently be augmented with cumbersome procedural descriptions.

Here is an example. In English I might ask for "A ranked percent of total amount broken down by product type." This sounds like a straightforward report, but in fact it requires a two-pass procedure to achieve the desired results. Below is the 4GL procedure required to answer this question:

```
TABLE FILE PROD
SUM PCT.AMOUNT AND HOLD
BY PROD-TYPE
END
TABLE FILE HOLD
PRINT PROD-TYPE
BY HIGHEST AMOUNT
END
```

In this we must first create a temporary file to HOLD the intermediate percent totals, and sort them by PROD-TYPE. Finally, the intermediate file is sorted and printed.

3.2.3 Logic Programming

A more refined approach to generalized descriptive programming is found in PROLOG —logic programming—in which the execution of the definition of the relationship between inputs and outputs replaces the procedural imperatives of how to compute outputs from inputs. For the BASIC program listed above, we could write an equivalent logical description as follows:

```
x is the greater of (x x)
y is the greater of (x y) if x < y
x is the greater of (x y) if y < x
```

These three definitions, which are very close to PROLOG syntax, completely describe what it means to look at two numbers and tell which is greater, but it doesn't tell us how to go about it. Built into PROLOG is all the "how" that's needed to use facts and rules such as the above to compute the extensive set of definite clauses of predicate logic.

To keep things in perspective, I should observe that even COBOL and FORTRAN have descriptive components. A FORTRAN or PL/1 statement such as $y = x + 3$, for example, is a description of the value to be assigned to y, and not a sequence of steps for computing the value. In fact every language which is at a higher level than machine language is higher precisely because it allows more descriptive specifications. Similarly, PROLOG has a procedural, or imperative, component which allows such things as reading and writing files (although in many PROLOG programs such commands are unnecessary). Therefore, a programming language is referred to as **procedural** if the predominant mode of writing programs is imperative procedures, and **descriptive** if the predominant mode is declarative definitions.

Generally, the more descriptive a programming language is, the easier it is to build programs that work, and the more closely will those programs resemble a specification of the problem to be solved. For business people, the payoff implied by that statement is that highly descriptive languages yield programs which are not only much easier to write, but also easier to maintain. This results from the programmer's needing only to maintain a close equivalent to the problem specification, not the machine's procedure for executing it.

3.3 Fifth-Generation Computers

In their on-going commitment to fifth generation computing, the Japanese suggest strongly that the next generation of computers will be firmly based on an architecture consisting of millions or billions of parallel processors, rather than our present sequential Von Neumann architecture. It is interesting that the Japanese have chosen PROLOG as the standard programming language for their Fifth Generation project as it underscores another benefit of highly descriptive programming languages: independence of machine architecture.

The imperative sequential structure of COBOL and FORTRAN is ideally suited to today's sequential machines, which put rather rigid constraints on the control logic of a program. Parallel processing architectures, on the other hand, tend to have processes driven by the data they need to transform. This is an ideal situation not only for PROLOG's declarative programs, but also for the DFD approach to system specification, where the mini-systems are intended to be independent parallel processors driven by their data needs.

3.4 The Downside of Description

When choosing system development techniques and protocols, we are constantly trading off between the ease of system development and the flexibility we have in the kinds of system we can develop. Every higher level language achieves its desirable "what" features by making some often sweeping assumptions about the kinds of problems you want to solve, and about "how" to implement those solutions.

In assembly language, for example, which is the closest approximation to machine language, we have complete freedom to implement any system the machine is capable of running, and doing it in any way we want! Unfortunately, assembly language requires also that the programmer, almost literally, keep track of everything: registers, buffers, print controllers, etc. This sounds much like the Turing Tarpit in which *everything* is possible, and *nothing* is easy!

PROLOG is no exception to the rule that higher level languages make implementation assumptions. In fact PROLOG has a rigid protocol for evaluating programs which is closely akin to the backward chaining inference technique described in Chapter 7, Inference and Knowledge. This implies a further assumption—that a problem can be specified in the language of predicate logic. The predicate logic assumption is not very restrictive at all and, in fact, most commercial applications fall easily into that paradigm. The assumption about backward chaining may be less generally applicable and for this reason I urge companies to *experiment* with PROLOG and other tools, but do not recommend standardizing on PROLOG at this time.

4. PROLOG for Prototyping

In our first pass at Generic Enterprises's knowledge-based Scheduler in Chapter 5, we stayed with the traditional paper models of structured analysis, although it was interesting to imagine some knowledge base prototyping in a few places. Now let's look at a few parts of the Scheduler analysis again, this time focussing on yet another approach to prototyping: programming in PROLOG.

This discussion introduces PROLOG capabilities assuming that you know little or nothing about the language. However, my goal is to show how PROLOG can be used effectively as a prototyping medium, and not to give you a flashy demonstration or to provide a PROLOG tutorial.[9] As a consequence, I shall not do the Scheduler completely, and many of the examples may seem somewhat oversimplified; this is intentional.

4.1 PROLOG

PROLOG is a descriptive specification language. PROLOG solves a broad class of problems. PROLOG is easy to learn. PROLOG is part and parcel of the fifth generation!

In many ways, PROLOG seems like the answer to a long-standing prayer for help in moving swiftly from problem perception, to specification, to working system. However, things are not quite so rosy as that; I'll point out PROLOG's shortcomings as we go.

4.1.1. General Characteristics

Essential standard syntax PROLOG is a cryptic, barebones language in which a "program" is the statement of a conclusion followed by conditions under which that conclusion holds true. In logic programming, all computation is a process of logical inference, and so we can see right away that it has potential for expert systems work.

There are many languages of logic, some more powerful in their expressiveness than others. PROLOG is a compromise between the most expressive forms of logic and forms that can be efficiently implemented on today's computers. PROLOG uses a very straightforward clausal form of logic in which each clause is an assertion of the form:

<This One Conclusion is true> if <a Conjuction of conditions is true>

Although people do write programs in the PROLOG standard syntax, it is common to find front-ends to the standard syntax which extend the language and make using it more programmer-friendly. For the examples used in this book I have used micro-

[9] Several good PROLOG tutorials are now available. I recommend Clark, K.L. & McCabe, F.G., *micro-PROLOG: Programming in Logic,* Prentice-Hall International, Englewood Cliffs, N.J., 1984 as a fairly clear introductory tutorial. Also good is the tutorial part of Borland International's *Turbo Prolog; Owner's Handbook,* Scotts Valley, California, 1986.

PROLOG and a front-end to it called apes, Augmented PROLOG for Expert Systems[10] on an IBM PC with two floppy discs and 256K of RAM. Although it is possible to use PROLOG in such a modest environment, clearly one cannot do very interesting problems without substantially more memory, disc space, and computer speed.

Following are a few examples of the difference between apes and standard syntax micro-PROLOG.[11]

apes	Standard
Mark likes Mary	((likes Mark Mary))
Mary likes Mark	((likes Mary Mark))
x is-a-friend-of y if	((is-a-friend-of x y)
x likes y and	(likes x y)
y likes x	(likes y x))

In general, front-end extensions to standard PROLOG allow considerable flexibility in the way clauses are stated.

4.1.2 A Programming Example

The first two of the above statements are facts which would be stored as data in PROLOG's knowledge base, and the third is a rule stating a relationship among facts. Rules such as is-a-friend-of are stored in PROLOG's knowledge base along with data and are the essence of a PROLOG program. To execute the is-a-friend-of program, we ask PROLOG either to find facts for which the rule is true or, alternatively, to confirm that at least one set of such facts exist without actually listing them. These two cases are illustrated below.

(1) the command

```
find(x and y: x is-a-friend-of y)
```

will produce a list of all pairs of people in the knowledge base for which the above rule is true. This statement can be read as follows: "Find all the pairs of x and y for which it's true that x is a friend of y, and print them out in the format 'x and y.'"
In this case the answers would be:

```
==> Mark and Mary
==> Mary and Mark
```

[10] This particular micro-PROLOG was developed by Logic Programming Associates, 10 Burntwood Close, LONDON SW18 3JU England. Another PROLOG interpreter and expert system extension, ES/P, is marketed by Expert Systems International, whose U.S. headquarters is in King of Prussia, Pa. At this time at least a dozen PROLOGs are available in the U.S. for micros and mainframes alike.

[11] Standard Edinburgh PROLOG, the original PROLOG implementation, has a somewhat different syntax from standard micro-PROLOG, but the primitive flavor of the languages is similar. My examples may not be completely accurate for the syntax of any particular PROLOG system, although I have tested the logic of every example using micro-PROLOG.

At this point don't be concerned about the meaningless redundancy of this answer. It is literally correct given micro-PROLOG's inference strategy, and there are features in the language which can be used to reduce such redundancy when it's appropriate.

We could also have filled in data for some of the variables, x y, and said:

```
find(x:x is-a-friend-of Mary)
or find(x:Mary is-a-friend-of x)
or others.
```

(2) the command

```
confirm(x is-a-friend-of y)
```

would yield the answer

```
==> YES
```

which means that for at least one pair of individuals the is-a-friend-of relationship holds true.

Once again, we can fill in some of the variables, x y, with data and phrase questions such as:

```
   confirm(Mark is-a-friend-of x)
or confirm(x is-a-friend-of Mark)
or confirm(Mary is-a-friend-of Mark)
```

Even with this very simple micro-PROLOG program, we have multiple levels of inference. That is, the relationship *is-a-friend-of* cannot stand on its own—it must have another relationship, *likes,* as part of its definition. The possibility for networking relationships in micro-PROLOG programs is vast and provides the machinery we need to use micro-PROLOG in prototyping DFDs.

4.1.3 Beyond Program Execution

Another reason for using a front-end such as apes is to have such things as an explanation facility, which gives an explanation of why/how micro-PROLOG reached a certain conclusion. Using apes, once we have asked *find(x and y:x is-a-friend-of y)* and received an answer such as *Mark and Mary,* we could ask why and get the following result:

```
To deduce
      Mark is-a-friend-of Mary
I used the rule
x is-a-friend-of y if
```

```
                    x likes y and
                    y likes x

        I can show
            1            Mark likes Mary
            2            Mary likes Mark
```

This is an extremely useful facility because, even in the simplest of prototypes, it is easy to forget the details of the rules being used, and it's a blessing if the system can remind you.

4.1.4 Knowledge Representation

In the examples in the following discussion, the knowledge is very simple and generally unstructured. Do not be misled by this apparent lack of sophistication; as we see in Chapter 8, Frames: Data, Knowledge, and More, micro-PROLOG can easily handle the most complex knowledge structures in use today.

4.2 Structured Analysis and Descriptive Specification

We are going to return now to the intelligent Scheduler system and rethink what we have done during structured analysis in terms of our stated goal, which was to provide a descriptive specification of the Scheduler. For a few of the mini-systems, we shall produce working prototype processes using micro-PROLOG. Please refer to DFD Figure 5, Develop Schedule, shown as Figure 6-2.

I previously explained that each of these mini-systems—Generate-Suitability-List, for example—is a domain-specific inference engine and that the process itself contained no rules, but only the inference logic for using rules in the knowledge base. Eventually, I drew a DFD (Figure 5-7) for Generate-Suitability-List which showed both the knowledge structure and the inference processor. I suggested that because Generate-Suitability-List had its own inference needs, we needed to show this level of detail.

Now I am going to change some basic assumptions. In fact, the foregoing is only one among many ways of representing this system, and certainly not the only approach to using DFDs for knowledge-based systems. In Chapter 4 I indicated that sometimes it might be appropriate to show in DFD form the knowledge base structure only, and not the inference engine. This, I said, was likely to be the case if we were using a general-purpose global inference engine which applied to all mini-expert systems. PROLOG is such a global inference engine.

In Figure 6-3 I have redrawn DFD Figure 5, with two notable changes:

(1) Process 5.2, Generate-Suitability-List, has been broken into three simpler pieces:

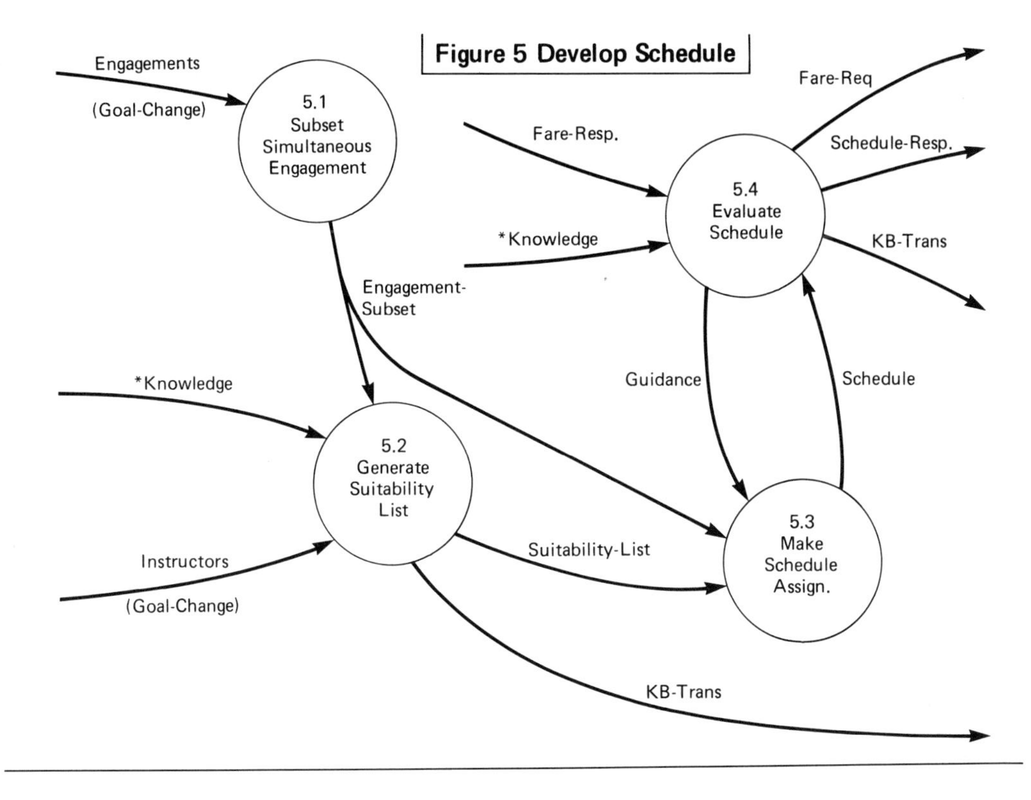

Figure 6-2. DFD Develop-Schedule Processing

Merge-Engagement-and-Instructor, Assign-Suitability-No, and Filter-DeFacto-Assignments.

(2) We no longer see Knowledge as an input to *any* mini-system! This is the more dramatic change, and it implies a fundamental change of attitude. That is, we no longer look at the mini-systems as inference processors, but rather as **descriptive specifications** of the knowledge required to transform its inputs into its outputs.

Even though this view limits our processing potential to the capabilities of the assumed inference engine, it is much truer to the problem-specification purpose of DFDs and makes it easier to think about micro-PROLOG as a prototyping medium for our DFD. Micro-PROLOG is now the inference engine for *every* mini-system, and we are considering *every* mini-system, even the almost trivial Engagement-Subset mini-system, to be an expert system knowledge base in its own right.

If micro-PROLOG turns out to be inadequate for the production version of the scheduler system, we may have to come back to our original view of the system as a highly structured knowledge base plus many domain-specific inference processors. If we come

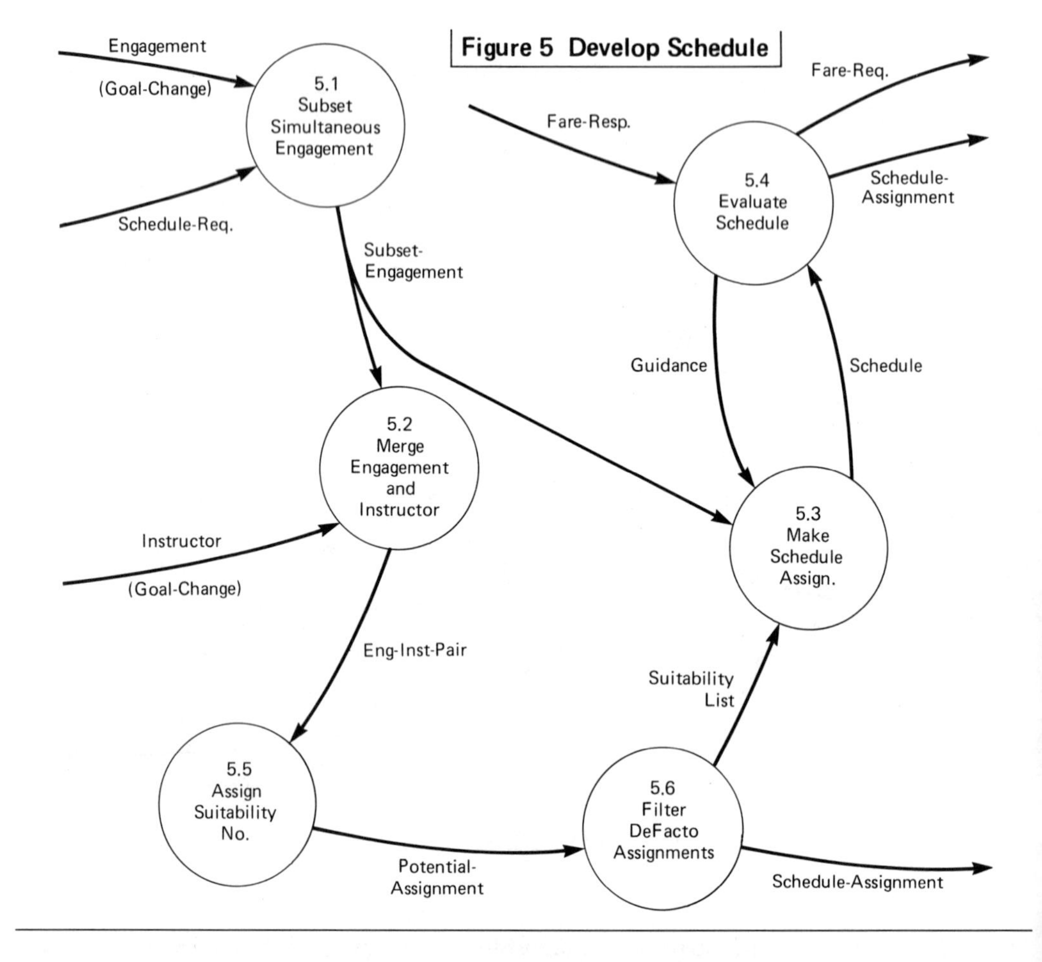

Figure 6-3. DFD Develop-Schedule Processing

to this, however, we shall see that the DFDs showing the inference processors almost exactly mirror the structure of the knowledge base DFDs. The result of this parallelism is that we may write multiple mini-specs for many of the mini-systems in order to specify both the inference processor and the descriptive knowledge components; this was illustrated in Chapter 5, Figure 5-7. The way this happens in practice will vary tremendously from situation to situation and won't be discussed in detail here.

4.3 Prototyping Develop-Schedule

When prototyping, even though some simplifying assumptions may be made, we still try to build a system one step at a time. This means that we build one piece, make it work alone, then build a second working piece, and then make those two work together.

Thereafter the process is one of adding a single working piece at a time until the final result is achieved. For the Scheduler, our model for incremental construction is the new DFD Figure 5, Develop-Schedule. We shall work with the new DFD Figure 5 of text Figure 6-3 and illustrate the development of PROLOG prototypes for mini-systems 5.1 (Subset-Simultaneous-Engagements), 5.2 (Merge-Engagement-and-Instructor), 5.5 (Assign-Suitability-No), and 5.6 (Filter-DeFacto-Assignments).

One can actually start anywhere since all the pieces are interrelated, but it is generally a good idea to take a good cut at defining the structure of the knowledge we're working with. In fact this job should evolve as the DFDs evolve. Although PROLOG is capable of using complex structures, I have chosen to be *very* simple for this prototype. Figure 6-4 is a listing of not only the data dictionary for Develop-Schedule, but the entire database as well. Obviously this is simpler than the information needed by the suitability decision structure developed in Chapter 5 and shown as Figure 5-3.

4.3.1 Subset-Simultaneous-Engagements

When using micro-PROLOG as a prototyping tool, our goal is to state a relationship between inputs and outputs in such a way that when the conditions of the relationship are met, then the appropriate output data will be available. Typically, the relationship names will be close to the name of one or more output data.

In the case of mini-system 5.1, Subset-Simultaneous-Engagements, I have chosen the relation name *Subset-Engagement,* so mini-spec 5.1 will have the form:

```
x Subset-Engagement if <certain conditions are true>
```

The conditions we need are quite straightforward being that the seminar dates are within the week we are concerned with. This is almost a complete statement of the "program" for Subset-Simultaneous-Engagements and is very close to what we can write in apes micro-PROLOG:

```
x Subset-Engagement if
   Schedule-Req (y z)
& (x1 x2 x3 x4)   Engagement
& (x2 x3) overlaps (y z)
```

There are just a couple of issues to deal with: How does apes know what week we are concerned with, and how does it know what *overlaps* means?

Built into the apes front-end is an awareness that if a relationship is not defined by at least one sentence, then that relationship is automatically "interactive." This means that when apes encounters an undefined relationship, it will ask the user. In our case, with no other definitions, apes would ask about both *Schedule-Req* and *overlaps.* Apes does know about *Engagement* from our database, however, and in executing this program, it would consider each engagement as a candidate for Subset-Engagement.

```
Engagement = Customer + Start-Date + End-Date + Seminar

           (MMS     850506    850510 PDQ)
           (GEX     850506    850508 AI3)
           (RCO     850508    850510 AI3)
           (IBS     850513    850517 SRD)

Eng-Inst-Pair = *The combination of one engagement and one in-
                 structor. *
              = Engagement + Instructor-Name

Instructor = Instructor-Name

           Keller
           Benson
           Zells
           Yourdon
           Foobong
           Chauncey

Potential-Assignment = *A combination of engagement, instructor,
                        and suitability number. *
                     = Engagement + Suitability-No + Instructor

Schedule-Assignment = *A firm combination of one engagement and
                       one instructor. *
                    = Eng-Inst-Pair

Schedule-Req = *A request from the Scheduler to develop a schedule
                for a particular week. *
             = Subset-Monday + Subset-Friday

Subset-Engagement = *One engagement which falls within a particu-
                     lar week *
                  = Engagement
```

Figure 6-4. Develop-Schedule Data Dictionary

Because we want the human scheduler to be in control of the scheduling process, it's likely that we will leave *Schedule-Req* undefined for apes, so that the human scheduler will always be asked for Monday and Friday of the subset week. On the other hand, we might as well have apes do the burdensome clerical task of testing overlapping engagements. To do this we define another relationship, *overlaps,* as follows:

```
(x y) overlaps (X Y) if
      (z1 x y z2)   Engagement
  &     (either x is-on-or-before X
            & y is-on-or-after  X
        or    x is-after X
            & x is-on-or-before Y)
```

This is a fairly cryptic statement, and in current micro-PROLOG it is possible to use variable names that are more interesting than variations on x, y, and z. So in this micro-PROLOG we might define *overlap* as follows:

```
(Start-Date End-Date) overlaps (Subset-Monday Subset-Friday)
 if (Customer Start-Date End-Date Seminar)  Engagement
  & (either Start-Date is-on-or-before Subset-Monday
          & End-Date is-on-or-after Subset-Monday
     or     Start-Date is after Subset-Monday
          & Start-Date is-on-or-before Subset-Friday)
```

This is clearly a more readable statement of the definition, although I shall have occasion to use both styles. It does point out yet another need for definition, and that is the relations *is-on-or-before* etc. We can do this with two definitions:

```
x is-on-or-before x

x is-on-or-before y
if x LESS y
```

The chain of lower- and lower-level definitions continues until something is defined in terms of primitive concepts which micro-PROLOG understands natively. In this case, the concept LESS is such a primitive, and the chain of subordinate definitions ends here.

So we now have a PROLOG program definition, or group of definitions, which is a prototype of Process 5.1, Subset-Simultaneous-Engagements, and which works on our database as follows:

```
find(x: x Subset-Engagement)
```

This will ask one question about the Monday and Friday of the subset week to get the Schedule-Req input, and produce three answers as a result of filtering all of our Engagements through its Subset-Engagement conditions. It produces each answer independently, in keeping with the philosophy of DFDs in which all data are passed along the dataflow, but usually one piece at a time. After each answer we must tell the system what to do next. One choice is to say "more" in which case it will produce the next answer.

```
Please list Monday and Friday, (x y), of the subset week:
```

```
(850506 850510)

==> (MMS 850506 850510 PDQ) .more
==> (GEX 850506 850508 AI3) .more
==> (RCO 850508 850510 AI3) .more

No (more) Answers
```

It may seem fairly useless to have apes print out these answers since the Subset-Engagements are just an intermediate step towards developing a schedule. As it happens, if we use the Subset-Engagement relation in another definition, rather than a printout, the values will become a part of that new relation when it is executed; in other words, referring to the dataflow Subset-Engagement causes all values of it to flow along the dataflow from mini-system 5.1 to 5.2 or 5.3 one at a time, exactly as we would wish.

The printout in this example allows us to verify that the Subset-Engagement mini-system works alone on this small database and can be trusted when used in other definitions.

4.3.2 Merge-Engagement-and-Instructor

Now that we have developed a set of those engagements occurring in one week only, we need to use those Subset-Engagements in mini-system 5.2, Merge-Engagement-and-Instructor, to create a set of records that are output one at a time, in which each Engagement is paired with each Instructor. The job here is also straightforward, and we are introduced to a new micro-PROLOG primitive relationship, APPEND.

APPEND is the main relationship used by micro-PROLOG to manipulate lists of items. Here we are trying to take a list of items which represent a Subset-Engagement (y) and APPEND to that an item which is an Instructor-Name (z) to give us a new list of items (x), which is the Eng-Inst-Pair. The definition follows:

```
x Eng-Inst-Pair
    if y Subset-Engagement
     & z Instructor
     & APPEND(y (z) x)
```

It is perhaps more clear from this example than the previous ones that micro-PROLOG's "conditions" are not always simply passive checkers of the truth of something. They frequently do real data transformations, such as APPENDing y and (z) to give x, and micro-PROLOG's success in doing that transformation determines the truth of the condition. (If we chose to do so, instead of just passing x on to another mini-system or printing it out, we could show a dataflow to a file and thus add our new Eng-Inst-Pair as a new fact in the database. For this system, however, merely passing it on as a temporary dataflow is adequate.) In addition to doing more than checking for truth, relations may also have side effects such as reading and writing files of data. These capabilities are illustrated in Chapter 8: Frames: Data, Knowledge, and More.

Running this mini-system will produce 18 answers, one for each combination of Subset-Engagement and Instructor. The answer to the query:

```
find(x:x Eng-Inst-Pair)
```

is perhaps a good place to look at the way micro-PROLOG evaluates queries.

```
==> (MMS 850506 850510 PDQ Keller) .more
==> (MMS 850506 850510 PDQ Benson) .more
==> (MMS 850506 850510 PDQ Zells) .more
==> (MMS 850506 850510 PDQ Yourdon) .more
==> (MMS 850506 850510 PDQ Foobong) .more
==> (MMS 850506 850510 PDQ Chauncey) .more
==> (GEX 850506 850508 AI3 Keller) .more
==> (GEX 850506 850508 AI3 Benson) .more
==> (GEX 850506 850508 AI3 Zells) .more
==> (GEX 850506 850508 AI3 Yourdon) .more
==> (GEX 850506 850508 AI3 Foobong) .more
==> (GEX 850506 850508 AI3 Chauncey) .more
==> (RCO 850508 850510 AI3 Keller) .more
==> (RCO 850508 850510 AI3 Benson) .more
==> (RCO 850508 850510 AI3 Zells) .more
==> (RCO 850508 850510 AI3 Yourdon) .more
==> (RCO 850508 850510 AI3 Foobong) .more
==> (RCO 850508 850510 AI3 Chauncey) .more
```

Micro-PROLOG evaluates the conditions in a definition in a left-to-right fashion. If it gets an answer to the first condition—in this case the Subset-Engagement (MMS 850506 850510 PDQ), which it finds by invoking the Subset-Engagement relation—it assigns this data to y and moves one condition to the right. It now tries to find something in the knowledge base which will satisfy this second condition. For our example, this is the first Instructor-Name, Keller, which it assigns to the variable named z. This second success sends it on to the third condition, where it attempts to APPEND (MMS 850506 850510 PDQ) to (Keller) and assigns the result to x as (MMS 850506 850510 PDQ Keller).

At this point there are no further conditions to the right of APPEND and no other ways to satisfy the APPEND relationship using the data it has in hand, and so it backs up one condition to the left and tries to find another value for Instructor. It finds Benson, assigns this to z, and once again moves on to the APPEND which yields (MMS 850506 850510 PDQ Benson) as the second solution. Micro-PROLOG will continue in this way until all Instructors have been exhausted for the first Subset-Engagement. It will then back up to the first condition to get another Subset-Engagement and process all the

Instructors and APPENDs for it. When all Subset-Engagements have been exhausted, the execution of this program is finished.

This left-to-right ordering of evaluations is PROLOG's procedural component, and it is capable of expressing any computation which can be represented in this form of clausal logic.

4.3.3 Assign-Suitability-No

The two preceding mini-systems might be thought of as being **data-intensive** in that our work was focused on transforming data according to single well-defined relations. In mini-system 5.5, Assign-Suitability-No, we become somewhat more **knowledge-intensive** in that the Suitability-No assigned to any Eng-Inst-Pair may be the result of applying multiple rules related to suitability.

Figure 6-5 lists a simple set of four rules for assigning a Suitability-No to certain Eng-Inst-Pairs. This simple list is far from realistic in its completeness, but it illustrates the use of rules in micro-PROLOG and a simple way of dealing with uncertainty.

Each of these rules defines an instance of the relation *Suitability-No-of* in which a particular Suitability-No is the result of certain conditions being true, and some are more

```
Rule 1:
90 Suitability-No-of (Cust Start-Date End-Date Seminar Instruc-
tor)
      if Seminar EQ AI3
       & Instructor EQ Keller

Rule 2:
0 Suitability-No-of (Cust Start-Date End-Date Seminar Instructor)
      if Cust EQ MMS
       & Instructor EQ Foobong

Rule 3:
40 Suitability-No-of (Cust Start-Date End-Date Seminar Instruc-
tor)
      if Seminar EQ AI3
       & not Instructor EQ Keller

Rule 4:
55 Suitability-No-of (Cust Start-Date End-Date Seminar Instruc-
tor)
      if Cust EQ GEX
       & Instructor EQ Keller
```

Figure 6-5. Assign-Suitability-No Rules

general than others. Rule 2, for example, assigns a Suitability-No to a very specific instance of one instructor and one company. Rule 3, on the other hand, assigns 40 as the Suitability-No to all AI3 seminars where Keller is not the Instructor. Clearly if we had included more data about locations, preferences, and other suitability items discussed in Chapter 5, we could easily evolve a much richer set of rules.

Presenting these rules first is backward, since we still have no relation which uses the Suitability-No. The following is such a relation.

```
x Potential-Assignment
  if (x1 x2 x3 x4 x5) Eng-Inst-Pair
   & y Suitability-No-of (x1 x2 x3 x4 x5)
   & APPEND((x1 x2 x3 x4 y) (x5) x)
```

The effect of this is to get an Eng-Inst-Pair from the input dataflow and use the applicable Suitability-No-of rule on it and then APPEND Cust Start-Date End-Date Seminar and Suitability-No followed by the Instructor-Name. Using this definition with the rules above, we would get the following answers:

```
==> (MMS 850506 850510 PDQ 0 Foobong) .more
==> (GEX 850506 850508 AI3 90 Keller) .more
==> (GEX 850506 850508 AI3 55 Keller) .more
==> (GEX 850506 850508 AI3 40 Benson) .more
==> (GEX 850506 850508 AI3 40 Zells) .more
==> (GEX 850506 850508 AI3 40 Yourdon) .more
==> (GEX 850506 850508 AI3 40 Foobong) .more
==> (GEX 850506 850508 AI3 40 Chauncey) .more
==> (RCO 850508 850510 AI3 90 Keller) .more
==> (RCO 850508 850510 AI3 40 Benson) .more
==> (RCO 850508 850510 AI3 40 Zells) .more
==> (RCO 850508 850510 AI3 40 Yourdon) .more
==> (RCO 850508 850510 AI3 40 Foobong) .more
==> (RCO 850508 850510 AI3 40 Chauncey) .more

No (more) Answers
```

If you look carefully at these answers you will find two problems, both of which are solvable although we may have to be a bit "cute": one is that several Eng-Inst-Pairs have been left out, and the other is that GEX's AI3 seminar has *two* Suitability-Nos for Keller. What's up?

The first of these issues, missing output, points up a serious limitation in PROLOG's approach to logical inference. That is, unless there are rules which will specifically get PROLOG to an answer, it won't find one. For example, none of the four rules listed in Figure 6-5 contains a combination of conditions which will assign a Suitability-No to

(MMS 850506 850510 PDQ Keller). Hence no Suitability-No will be assigned to that item and it will be left out of the answer list.

This is a problem not only with PROLOG, but with nearly all of the current rule-based expert system shells, whether they cost $95 or $40,000. The only systems that avoid this problem are those which use, in addition to rules, a kind of analogical inference as does the TIMM system from General Research Corp.

The second problem, multiple outputs, occurs because there are two rules which yield different Suitability-Nos for the GEX AI3 seminar taught by Keller, and here we encounter the need to deal with uncertainty in the knowledge base.

Both of these problems are "solved" by adding one new Suitability-No-of rule:

```
99 Suitability-No-of (Cust Start-Date End-Date Seminar
                                Instructor)
```

and by changing the *Potential-Assignment* relation so that for any one Eng-Inst-Pair it will use the minimum Suitability-No-of value. Missing data is solved because our new rule forces a 99 Suitability-No-of to be assigned to every Eng-Inst-Pair, and multiple Suitability-Nos is solved by choosing the smallest of those returned. It can be done as follows:

```
x Potential-Assignment
 if (x1 x2 x3 x4 x5) Eng-Inst-Pair
  & y1 ISALL(y:y Suitability-No-of (x1 x2 x3 x4 x5))
  & y2 minimum-of y1
  & APPEND((x1 x2 x3 x4 y2) (x5) x)
```

There are three changes here:

(1) We introduce another micro-PROLOG relation called ISALL, which in this case takes all of the answers to repeatedly applying Suitability-No-of rules to one Eng-Inst-Pair and wraps them up in a list which is assigned to y1.

(2) With this list of all the Suitability-Nos, we can use a simple minimum-of relation to find the lowest one and assign it to y2.

(3) y2 is now used in the APPEND relation to make one and only one Potential-Assignment to x for each Eng-Inst-Pair.

This is definitely not a very elaborate scheme for dealing with uncertainty, but it works for this small example. (Although Chapter 7, Inference and Knowledge, addresses the issue of uncertainty further, the details of uncertainty in expert systems are beyond the scope of this book.)

Following is the list of answers using the new *Potential-Assignment* relation and the 99 Suitability-No-of rule:

```
==> (MMS 850506 850510 PDQ 99 Keller) .more
==> (MMS 850506 850510 PDQ 99 Benson) .more
==> (MMS 850506 850510 PDQ 99 Zells) .more
```

```
==> (MMS 850506 850510 PDQ 99 Yourdon) .more
==> (MMS 850506 850510 PDQ 0 Foobong) .more
==> (MMS 850506 850510 PDQ 99 Chauncey) .more
==> (GEX 850506 850508 AI3 55 Keller) .more
==> (GEX 850506 850508 AI3 40 Benson) .more
==> (GEX 850506 850508 AI3 40 Zells) .more
==> (GEX 850506 850508 AI3 40 Yourdon) .more
==> (GEX 850506 850508 AI3 40 Foobong) .more
==> (GEX 850506 850508 AI3 40 Chauncey) .more
==> (RCO 850508 850510 AI3 90 Keller) .more
==> (RCO 850508 850510 AI3 40 Benson) .more
==> (RCO 850508 850510 AI3 40 Zells) .more
==> (RCO 850508 850510 AI3 40 Yourdon) .more
==> (RCO 850508 850510 AI3 40 Foobong) .more
==> (RCO 850508 850510 AI3 40 Chauncey) .more

No (more) Answers.
```

4.3.4 Filter-DeFacto-Assignments

Let's look at just one more step in the prototyping process: the removal of defacto assignments and the generation of the Suitability-List which is input to the Make-Schedule-Assignments mini-system. This presents an interesting new problem of producing two outputs from a single Potential-Assignment input. Furthermore, we would now like the Schedule-Assignment output for defacto assignments to be added to the database permanently.

As with most descriptive solutions, it is best to start at a high level, and fairly simply. Let's make the following definition:

```
x Suitability-List
  if x ISALL(y:y Potential-Assignment
                & not y defacto
                & not 0 Suitability y)
```

This tells micro-PROLOG simply that we want a list of all the Potential-Assignments which are not defacto assignments and which have a Suitability other than 0. All we have to do now is tell PROLOG what we mean by *defacto* and *Suitability*.

In this case, Suitability is simply the fifth element in each Potential-Assignment, and we can define this as:

```
x5 Suitability (x1 x2 x3 x4 x5 x6)
      if (x1 x2 x3 x4 x5 x6) Potential-Assignment
```

Defacto is more difficult, since we have to look at all the Potential-Assignments for each Engagement, and if there is only one possible, or one with non-zero Suitability, then it is a defacto assignment. The following definition works, although it requires some fairly heavy computation from micro-PROLOG:

```
x defacto
     if y ISALL(z:z Potential-Assignment
                   & front(4 X x)
                   & front(4 Z z)
                   & X EQ Z
                   & not 0 Suitability z)
      & y HAS-LENGTH 1
      & (x Schedule-Assignment) ADD
```

What we are doing here is creating a list of all Potential-Assignments which are the same as the one we are checking, x, and which also have a Suitability other than 0. We then check to see if the length of that list is 1, and if so we add x to the database as a Schedule-Assignment.

5. Conclusion

In this chapter, we have seen that there are now many choices of environment for getting automated help with system development—either clerical help, or help in actually implementing prototype versions of the system. The cost of the different tools varies from a few hundred dollars for an interesting micro-PROLOG on the IBM PC to $200,000 or more for a sophisticated knowledge engineering environment and an AI workstation.

We have seen also that the traditional techniques of structured analysis lend themselves very directly to rapid prototyping using PROLOG, one of the first implementations of logic programming.

The new AI techniques can be a boon to data processing shops, helping them solve more complex problems than has been possible previously, or they can lead into a mire of unmanageable knowledge complexity. By continuing to manage AI software projects with the same care we have applied to traditional projects, the new can be blended with the old to yield high-payoff applications on time, on spec, and on budget.

Chapter 7
Inference and Knowledge

We have seen that PROLOG can be a powerful tool for rapidly prototyping systems in conjunction with structured analysis. A question which often follows a demonstration of PROLOG prototyping is: "What I really want to do is build an expert system using [one of the commercial shells]. So why, if I'm going to use a shell, do I care about either structured analysis or PROLOG?" The answer is twofold and simple. First, as I have pointed out before, *any* AI/expert system is still a system development project which strongly suggests the need for analysis—structured or otherwise—up front. Second, PROLOG *is* a shell, and a very competent and powerful one as we shall continue to see.

To reinforce: Whether you use a simple commercial shell or write LISP code and use an AI workstation to do your knowledge-based system, you are embarking on a system development project, and it should be treated as such.

To build a realistic expert system, we need to study some techniques that AI researchers have delineated and in so doing we shall look more carefully at the uses of PROLOG. In this chapter and the following, we shall focus on expert system structure and particularly those two specific areas which seem to be interestingly different from much of traditional programming: knowledge representation and inference. Both areas have been subjected to intensive research for many years, and the results so far are both useful and surprisingly easy to implement.

1. What is AI?

I keep returning to the question of what we mean by AI, and I never feel as though a completely adequate answer has been given. Possibly there is no satisfactory answer, but here is another attempt.

1.1 AI Means Expert System

I suggested earlier that nobody really knows what intelligence is, or how to say for sure when someone is behaving intelligently. The same is true of artificial intelligence, which has become such a catch-all term for a very fuzzy group of tools and techniques that it is almost meaningless.

A primary goal of AI research is to find general computational models for intelligent human processes. However, the more we look into the way people do things, the more it appears that intelligence is not based on having general models of the world, but rather is the result of skillfully collecting and applying large quantities of anecdotal information, that is to say heuristics. As we have seen, heuristics are the foundation of what we have been calling expert systems.

Thus it seems also that almost everything useful that's being done today in AI is heavily based on the collection and use of heuristics. So in a very real sense, AI means expert system. Even the natural English front-end systems are themselves expert systems whose expertise is in the grammar they use to understand an English statement.

With this in mind, the restricted focus of this book, knowledge-based expert systems, seems not so restricted. The principles of analytical representation we have discussed so far are applicable not only to the kinds of expert system we have addressed, but also to natural language understanding, robotics, and others. For AI in general, however, there is a vast array of differences in knowledge representation techniques, inference techniques, and control structures, and I have not begun to deal with this diversity.

1.2 Expert Systems

In general we have been addressing a class of AI system called **production systems,** so named because of their underlying dependence on condition-action rules, often called **production rules.** In a certain sense, this class of systems is a relatively small part of AI, but it happens to be a part in which most of AI's usefulness has been demonstrated. Focusing on a small part of the AI field has the advantage that we don't have to define concepts like *knowledge* or *thinking* or *intelligence* in general, but can deal with them in relatively predictable ways.

Earlier I referred to the definition of experts as "people who learn more and more about less and less until eventually they know everything about nothing." One gets the feeling that gaining expertise is a narrowing and focusing experience.

This is particularly true of expert systems, in which we address a very narrow area of human expertise and attempt to transfer to the computer all the knowledge about that one apparently tiny domain. Once the knowledge is in the computer, the inference engine uses that knowledge to reach decisions or make plans. As we discussed in the chapter on application selection, having the machine do a task in the often inefficient way a human does may not be the best choice, but it is an assumption implicit in the building of today's expert systems.

So what is knowledge in this restricted sense? What is inference, or reasoning? And how do we make it all happen in these intrinsically unintelligent computers?

2. Rules as Knowledge

Throughout this book I have used IF-THEN rules without explaining why they were being used or exactly how a machine would use them to exhibit something like intelligence. We must now look more closely at such rules, since they are a major tool in representing knowledge in virtually every expert system which has been done, is now being done, or is likely to be done for some time. Even those systems using a more elaborate frame representation[1] typically have one or more sets of rules for making decisions or executing plans.

2.1 What Kind of Knowledge?

Several decades ago, the well-known Swiss psychiatrist Carl Jung described a way of looking at types of human experience and the ways that we process that experience. The general categories of his psychological types are still a good first cut at describing the kinds of knowledge we would like to represent in a computer. Figure 7-1, Knowledge—A

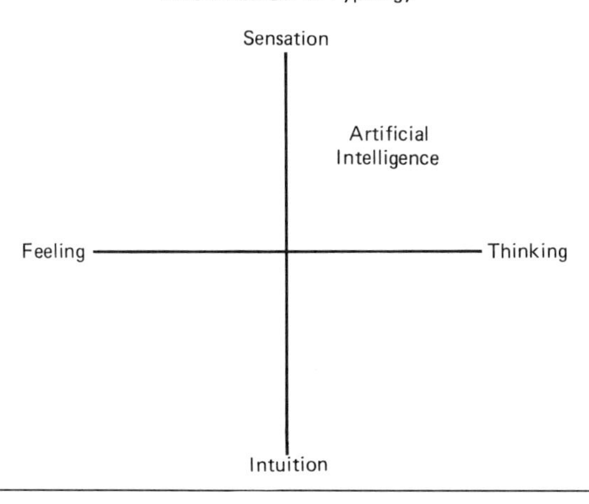

Figure 7-1. Knowledge—A Typology

[1] Frame knowledge representation is elaborated in Chapter 8.

Typology, illustrates the general elements of Jung's analysis.[2]

This simple presentation of Jung's types shows pairs of opposing functions which are available to process our experiences: sensation and intuition for perceiving experience, and thinking and feeling for evaluating experience. These functions are the ways that we know things.

The **sensation** function, implemented by our five physical senses, is what we use to perceive events in the physical world. These sensory perceptions are passed on to what Jung called the **thinking** function, whose job is to organize the sensory input according to one or more logical-deductive paradigms and possibly to hand that information to a **feeling** function. As Jung used the term, the job of the feeling function has little to do with emotional feelings, but rather addresses the task of judging whether or not something feels right or wrong, good or bad. Finally, the **intuitive** function is a kind of inner perceiving function. It appears to be able to extend past experience, allowing us to look beyond our present situation to see what may be coming next. The pairs of functions are shown as opposing poles in Figure 7-1 to indicate that one cannot simultaneously use the sensation and intuitive perceiving functions, or the thinking and feeling evaluative functions.

The importance of this diagram for our purposes is to indicate that the capabilities of AI—and in fact of all of computer science—are limited to the perception of physical inputs and their processing by logical-deductive thinking processes. We have been very successful at using logical IF-THEN rules to represent essentially logical expertise because that's what current computing machinery is good at.

This is not to say that we can't emulate the feeling and intuitive functions on present computers. In fact, much of the skill in building a satisfactory knowledge base has to do with discovering and codifying apparently non-logical expert activities using our logical-deductive tools. Often we find that intuition is simply a very clever application of logical-deductive techniques. More than this, the work of Roger Schank's AI research group at Yale University[3] has resulted in systems which impressively represent not only logical knowledge, but empathetic and interpersonal feeling-toned themes as well.

As impressive as these apparently non-logical results are, the machine is not feeling, or intuiting, etc. The systems are emulations of feelings, approximations which use well-defined, sequential, logical-deductive hardware and software processes.

[2] For a more thorough discussion, see Jung, C.G., *Psychological Types,* Harcourt Brace, N.Y., 1938, or the more recent *Collected Works of C.G. Jung,* Volume 5, Princeton University Press–Bollingen Series, Princeton N.J., 1960. Jung's work included many refinements of this basic typology, including extraversion/introversion. Subsequent work has enhanced Jung's ideas to include dimensions of judgment/perception and others. Although such elaborations are beyond our needs here, further reading can best be sought in the contemporary literature of organizational dynamics and personnel evaluation.

[3] Some of this work is discussed further in Dyer, M.G., *In-Depth Understanding,* Yale University, Research Report #219, May 1982.

2.2 What Rules Mean

Although I have been referring to IF-THEN or production rules throughout this book, I have not been very specific about what can go into such rules or the issues that arise as a result of using them.

When we speak of rules for representing knowledge, we typically mean the following:

```
IF some things are true
THEN some other things are true
```

For example, I could make the rule:

```
IF the temperature is less than 30 degrees
THEN put on ear muffs
```

A similar rule from the world of computers might look like this:

```
IF the input transaction is Add-a-Record
THEN execute the Add-a-Record process
```

"So what's new?" you say. "Haven't we always written this kind of rule in our systems?" Of course we have. Every programming language has some explicit or implicit equivalent of the statement:

```
IF some things are true
THEN some other things are true
ELSE yet other things are true
```

However, there is an all-important difference between traditional IF-THEN rules and the production rules used in knowledge-based systems: certainty, completeness, and dynamism.

Typically when we write IF-THEN rules in a COBOL program, we are confident that if the conditions are definitely true, then the conclusions are definitely true, i.e. if the input transaction is Add-a-Record, then we definitely want the Add-a-Record process to be executed. In such rules the conclusions are completely determined by the truth of the conditions (premises); these are called *deterministic* rules. Other rules may imply the same conclusion, but for this set of conditions we know the result without question.

Knowledge-based system rules tend to differ in that when we make an expert rule, we may not be absolutely certain that either the premises or the conclusions are true. In the ear muff example, I will not always put on ear muffs when the temperature goes below 30 degrees. It actually depends on many other conditions:

```
Whether I have a high collar on my coat.
Whether or not the sun is out.
Whether or not and how long I've been exercising.
How I feel about wearing ear muffs.
How long I expect to be outdoors.
The consequences of not putting on ear muffs.
Etc.
```

In addition, each of the new factors may be true only to a certain degree, and certainly that degree will vary from person to person. So if I'm trying to develop a knowledge base for outdoor expertise, the ear muff rule needs not only additional premises, but they may be stated in a fuzzy way. Further, the conclusion is true with only a limited certainty.

Here is a less well-defined statement of the ear muff rule indicating the kind of uncertainty which may be present:

```
IF I don't have a high collar (How high? What material?)
  & The sun is not out (How far out? What time of day?)
  & I haven't been exercising (How much? What kind?)
  & I don't like ear muffs (In what way? How much?)
  & I will be out a long time (How long? What shelter?)
  & The temperature is below 30 degrees (How far below?)
  & The alternative is pain (How much? Permanent damage?)
  & Etc
THEN Put on earmuffs (about 75% of the time)
```

Such rules are better described as guidelines or rules of thumb; the word we have been using for such rules is *heuristic.*[4] A heuristic is a bit of anecdotal information we have collected somewhere in our experience which can be applied when making decisions in future situations which are similar, but not quite the same. The heuristic doesn't tell us specifically what is right or wrong, but it helps us discover for ourselves which of two or more alternative paths may lead us to a solution in a new situation.

It appears that human expertise is heavily based on having large compilations of heuristics combined with some skill in knowing when and how to apply them. Knowledge-based expert systems typically contain from a few hundred to a few thousand heuristics and possess software—what we have been calling an *inference engine,*—which uses the heuristics in making complex decisions and in planning.

The problems inherent in using such heuristics to make decisions are manifold. For example, if I have another rule which uses the wearing of ear muffs as one of its premises, how do I use the certainty of the ear muff rule conclusion to derive a certainty for the second rule?

[4] From a Greek word "heuriskein," meaning to discover or invent by oneself, to find.

2.3 The Inference Problem

The inference process actually involves several different processes that must work together—these can be broadly grouped as rule retrieval, conflict resolution, and execution. Very simply, this means that we do three things every time we wish to execute a rule:

1. Identify the rule as relevant to the conditions (premises) of the problem situation. This step may actually result in finding more than one rule from the knowledge base. If so,

2. We must resolve the conflict among competing rules, the result being the selection of one rule. Now,

3. We can execute the rule. That is, we can reach the conclusion implied by its premises.

In the following discussion, I have treated the issues of conflict resolution very simplistically in order to present the basic reasoning techniques.

3. Reasoning by Chaining Rules

Just as rules are the most common form of storage for expert knowledge, *rule chaining* is the most common way of using rules to reach decisions. We shall examine the basic form of the two most popular protocols of rule chaining—forward chaining and backward chaining—and consider some necessary and useful extensions to these basic forms.

3.1 Forward Chaining

Figure 7-2 is a list of simple rules. Let's say this is our entire knowledge base and we wish to use these rules in decision making. In using the rules, we are doing the reasoning work of the inference engine; in a sense, we are the inference engine. Using the technique of forward, or data-driven, chaining to reach a conclusion, we approach the knowledge base as though we know absolutely nothing ahead of time, and our job is to discover something. We will typically be given some facts about a situation, although we may have to ask for them explicitly.

The elementary method of forward chaining arbitrarily starts with the first rule in the knowledge base and try to use it.[5] We can use this, or any, rule if all of its premises are known to be true ahead of time. That is, we must somewhere keep what I shall call a "True List" on which we store all the things known to be true. The True List is empty at the start of every decision, and we fill it in as the decision process proceeds.

[5] This is the first place I have sidestepped the problem of rule conflicts. More typically we would have to look for *all* rules which satisfy the current situation.

Knowledge Base

```
1. IF due date on-or-before today
   THEN payment is due
2. IF due date is after today
   THEN payment is NOT due
3. IF payment is due
     OR cash available is excessive
   THEN paying is recommended
4. IF paying is recommended
     AND NOT known if paying is prevented
   THEN ask the user if paying prevented
5. IF paying is recommended
     AND paying is NOT prevented
   THEN pay the bill
```

Figure 7-2. Knowledge Base
©Copyright 1985 Renaissance International Corporation

Clearly, to put anything on the True List to start with, we need a way to find out things that can't be learned as conclusions of some rule in the knowledge base. We do this by asking the user—or going to a database—for information whenever the truth of a premise is not known and cannot be determined from another rule.

Once we do use a rule, we add all of its conclusions to the True List and look for the next rule in the knowledge base which uses the first rule's conclusion as one of its premises. We now try to use this new rule. *This logical motion from one rule's conclusion to another rule's premise is the forward chain.*

Figure 7-3 shows the same small knowledge base, with four arrows added to indicate the order in which rules would be tried in forward chaining. It shows also the contents of the True List associated with each step:

1. Start at the beginning of the knowledge base and try to use the first rule. The True List is empty, and if we look through the rest of the knowledge base, we see that the premise of rule 1 can't be proved from any other rule. For this reason, we know we must ask the user if the due date is on-or-before today. Let's assume the user says "Yes." Because the user has told us something, we can add that information to the True List and we know that the premise of rule 1 is true. Thus, we can execute rule 1 and add its conclusion to the True List.

```
TRUE LIST
(Forward Chain)

1 due date is today
  payment is due
2 paying is recommended
3 paying NOT prevented
4 pay the bill
```

```
Knowledge Base

         1. IF due date on-or-before today
┌────────── THEN payment is due
│1       2. IF due date is after today
│            THEN payment is NOT due
└───────→ 3. IF payment is due
              OR cash available is excessive
┌────────── THEN paying is recommended
│2──────→ 4. IF paying is recommended
│              AND NOT known if paying is prevented
│         3┌─ THEN ask the user if paying prevented
│4───────→ 5. IF paying is recommended
              AND paying is NOT prevented
             THEN pay the bill
```

Figure 7-3. Forward Chaining.

2. We now look for a rule which has the conclusion of rule 1 as one of its premises. Rule 3 is such a rule and we chain forward trying next to use rule 3. We know that "payment is due" is true, and since the two premises of rule 3 are joined by the OR condition, we need know only one of them in order to use the rule. We use rule 3 and add its conclusion, "paying is recommended," to the True List.

We now have a conflict, since both rule 4 and rule 5 have the conclusion of rule 3 as a premise. We are going to be simplistic in resolving this conflict by saying that this forward chaining methodology requires that we follow all possible chains using the physically closest rule first.

3. Thus, our next forward chain is to rule 4, the next one to have "paying is recommended" as a premise. Both of the premises of rule 4 are true, since whether or not the user will prevent paying is NOT known. So we use rule 4 concluding by taking the action specified in its conclusion, "ask the user. . . ." Let's say the user answers "No," paying is NOT prevented. This can be added to the True List, and the user's answer becomes the conclusion of rule 4. Thus our next action is to chain forward to rule 5.

4. All of rule 5's premises are known to be true and so we can use rule 5, adding its conclusion to the True List. This could be our final conclusion since we can chain to no other rules from rule 5. However, the forward chaining protocol says that we have to follow all forward chains, of which there is a second, from rule 3 to rule 5. We already know that both of rule 5's premises are true, so following this chain gives us no new information, although in a different example it could lead to additional sets of conclusions.

Forward chaining is best used when we are trying to answer a question which gives us some data and asks us to find some conclusion. Such a question here might be: "What do I do with a bill whose due date is today and for which payment is not prevented?" One of the main problems with forward chaining is that it typically has no basis for choosing one path over another and thus may have to search the entire knowledge base before coming up with an answer. In addition, the questioning of the user is likely to be quite random, which in practice quickly becomes extremely annoying. This problem alone has driven some real systems to use another chaining method.

It should be evident also that the need to continually look through the knowledge base to find conclusions which appear as premises, or questions which can be asked of the user, is very wasteful. In practice of course, this simple forward chaining inference engine would be enhanced considerably. As the knowledge base was being built, for example, we would flag those premises which are askable of the user, and index the premises and conclusions with regard to the rules in which they appear.

Yet another inefficiency was encountered when we re-proved, via chain #4, the final conclusion that the bill should be paid. This is somewhat more complicated to head off without actually indexing every possible chain of logic, which in a large knowledge base is typically impossible.

Another problem which is true not only of forward chaining, but also of the backward chaining method, is that it can deal only with combinations of circumstances which have been written into the rules, and I can reach a conclusion only if a direct chain of rules will get me there.

3.2 Backward Chaining

Backward chaining is sometimes called goal-driven chaining, or sub-goaling. It differs from forward chaining mainly in that we start by assuming some conclusion to be true and then use the rules to try to prove it.

The method of backward chaining starts with some conclusion and proves that conclusion by proving the truth of each of its premises in a left-to-right, or top to bottom, order. To prove the truth of a premise, we look for a rule which has that premise as one

```
            TRUE LIST
       (Backward Chaining)

   1
   2
   3 due date on-or-before today
     payment is due
     paying is recommended
   4 paying is NOT prevented
     pay the bill


         Knowledge Base

    1. IF due date on-or-before today
  ┌──→ THEN payment is due
 2│ 2. IF due date is after today
  │    THEN payment is NOT due
  └─ 3. IF payment is due
         OR cash available is excessive
  ┌──→ THEN paying is recommended
 1└─ 5. IF paying is recommended
         AND paying is NOT prevented
       THEN pay the bill
```

Figure 7-4. Backward Chaining

of its conclusions. If such a rule is found, we chain backward to that rule and attempt to prove it by proving the truth of each of its premises in the same fashion. The process ends when there are no further premises to be proved.

If you are feeling that you have read this before, re-read Chapter 6, Section 4.3.2, which describes how PROLOG evaluates queries—it should be evident that this is exactly the description of backward chaining. PROLOG is a backward chaining inference engine with considerably more power than most commercial shells that use the same technique.

Returning to our bill-paying example, we have the following scenario, which is illustrated by Figure 7-4. It would follow most likely from a query such as "Under what circumstances do I pay the bill?" (I have removed rule 4 from our knowledge base since it unnecessarily complicates this explanation of simple backward chaining.):

1. Once again we need a True List, which is empty at the start. We begin by assuming the conclusion, "pay the bill," and try to prove its first premise first. This means finding a rule which has "paying is recommended" as one of its conclusions. This we find in rule 3.

2. We chain backward to rule 3 and now attempt to prove that its first premise is true. We do this by finding rule 1, which has "payment is due" as a conclusion.

3. In rule 1 we wish to prove its first and only premise to be true but can find no rule which has that as a conclusion. So, as with forward chaining, we must ask the user who says, let's assume, "Yes," the due date is on-or-before today.

Notice that even though we have gone through two backward chains, we have not yet put a single thing on the True List. This is typical of backward chaining, which seems constantly to be putting off actually determining truth. Now that we know about the due date, however, we can add quite a bit to the True List, as you see in Figure 7-4—we have proved both premises of rule 1 and the first premise of rule 5.

4. Still working backwards, the next task is to deal with the second premise of rule 5, whose truth is not known and which can be determined only by asking the user. If the user says "No" then we have also proved the conclusion we started out with.

Backward chaining is probably a good inference technique candidate when you can reasonably guess at what the conclusion might be. There are at least two good reasons for this, one being that the logic chain will be as short and direct as possible in your knowledge base, the other being that the user questioning is focused on the goal being proved. Most commercial shells which offer only one inference technique will offer backward chaining. If you have no idea of what the conclusion is in a given situation, however, then backward chaining is probably no more efficient than forward chaining.

4. The Tree Variation

Although the IF-THEN rule representation we have been using is quite common, the same information could be represented in other ways. Figure 7-5 shows part of our bill-paying knowledge base in the graphic form of an AND/OR tree.

This representation was derived directly from the rules of Figure 7-4 by starting with the conclusion as the root of the tree at the top of the figure. The next level is found by listing both of the premises of rule 5 joined by the logical connector "&". Below that, each branch is a premise of the rule whose conclusion is its parent node.

There are some important things to notice about this way of looking at our bill-paying knowledge base. One is that this tree represents all of the possible logical paths one can follow to reach the "pay the bill" conclusion. Another is that it is much easier for a human to follow the logic in this graphic form than in the IF-THEN form. This is why the more sophisticated knowledge engineering environments display their knowledge in forms very similar to this.

Another thing to notice is what it means to do chaining with such a representation. Backward chaining, for example, means simply starting at the top of the tree and working your way down to the left, until you get to the bottom. Notice that all the bottom points, the leaves, are the questions which can be asked of the user. Once you get to the bottom by following one branch, come back up until you get to a node where there is another choice, and then follow that new branch as left as possible to the bottom. The procedure repeats until there are no further branches to follow. Forward chaining works similarly, but starts at the bottom of the tree with the askable questions and proceeds toward the top.

It's easy to see how the IF-THEN rules can be transmitted to the computer since we are used to dealing with English-like strings of characters. It may not be quite so clear how the graphic tree would be represented. There are many ways, but one common way is to store the tree as a list of lists as shown in the inset of Figure 7-5.

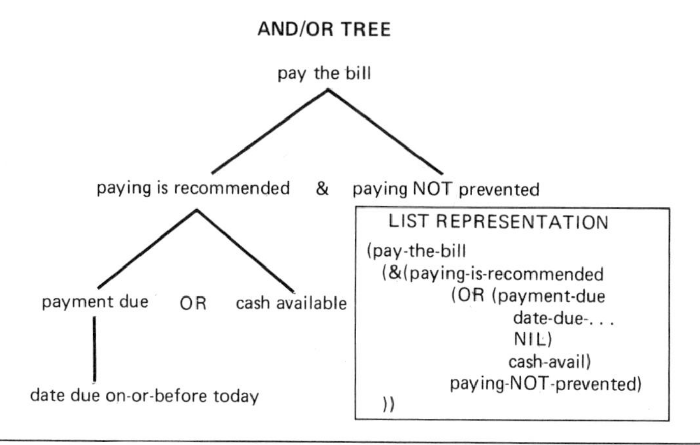

Figure 7-5. And/Or Tree

© 1985 Renaissance International Corporation

5. Enhancements to Rule Chaining

Working versions of the very simple rule chaining inference mechanisms I have described here could actually be coded on one or two pages. However, real inference engines may require hundreds or thousands of lines of code. This section considers some of the many possible enhancements which do appear in real systems, although at this early stage of the growth of AI they tend to be rudimentary capabilities.

5.1 Search Efficiency

We have mentioned already that there are inefficiencies inherent in the chaining techniques. Still, we can do some things at the time we build a knowledge base, as well as during execution, to remove some of the worst inefficiencies. I simply list them here, as they are very similar to what we would do to improve efficiency in a traditional data base situation:

- Cross-index premises and conclusions to the rules in which they appear.
- Mark those premises which do not appear as conclusions, and hence are askable questions.
- During execution, mark rules which have been used.

5.2 Conflict Resolution

Whether using forward or backward chaining or some other technique for working with rules, frequently situations arise in which more than one rule applies at any given time. This presents us with a conflict as to which rule to use, or which one to use first. The process of deciding issues such as the sequence of rule usage, where to focus the inference engine's attention, and what kinds of interrupts to allow, all fall under the heading of **conflict resolution.**[6]

In our example of forward chaining, we actually used two approaches to conflict resolution. First, at the very start, we arbitrarily chose the physically first rule in the knowledge base as our starting point. Second, when there was more than one chain of reasoning to follow, we decided to follow all of them in sequence starting with the one which is physically earliest in the knowledge base.

In spite of their simplicity, these are actually perfectly acceptable ways of resolving conflicts, *if they are truly applicable to the knowledge domain you're working with.* This is a *very* important condition, and one which will apply subsequently to our discussion of certainty factors. You can use any techniques that seem to fit, but typically they must be hand-

[6] Of the many texts which treat conflict resolution, two of particular interest are: Buchanan, B.G. & Shortliffe, E.H., *Rule Based Expert Systems,* Addison-Wesley, Reading Mass., 1984; Davis, R. & King, J.J., *An Overview of Production Systems,* in Elcock, E. & Michie, D. (Ed.), *Machine Intelligence 8,* Ellis Horwood, Chichester, England, 1977.

tailored to each individual domain based on comparisons of the system's performance against the human expert's performance.

This raises one of the major problems with commercial expert system shells as vehicles for production usage: they have built-in, hard-wired assumptions about inference technique, about conflict resolution, and about certainty. Despite outrageous advertising claims that you can use some shell to build any expert system, it is simply not true. As I stated before, if a specific shell product's assumptions work for your domain, or if there are "hooks" into the shell's code that allow you to change its assumptions, then the shell may be a good choice. On the other hand, if they don't provide a very close fit with the human expert's assumptions and methods, then you can probably use the shell for early prototyping and experimentation, but it's unlikely that you will ever be able to mature your system to a state of truly expert performance.

5.3 Certainty

In the example knowledge base of Figure 7-2 the rules, as stated, and the interaction with user, as described, are relatively well-defined; this is not typical.

In most cases, experts' knowledge has some kind of certainty associated with it, as does the input from a user at decision time. The important questions with regard to certainty of knowledge are: How do we decide how certain a particular piece of knowledge is *a priori*? and, How do we combine the certainties of uncertain premises to arrive at a reasonable certainty for the conclusion of a rule?

In some cases, it is possible to use a well-developed formal probability theory—something like Bayes' Theorem[7]—to arrive at the probability of a conclusion based on the probability of the premises. Unfortunately, the use of Bayes' Theorem implies some fairly restrictive constraints on the premises.[8] For one thing the premises must be mutually exclusive, and for another we must have prior probabilities available for the premises.

In fact some premises which appear to be mutually exclusive may not be. For example, let's consider the rule:

```
IF my son is going to be late for dinner
    and IF he can get to a phone
        THEN he will call
```

In this case, although the premises may sound unrelated, my son's getting to a phone is relevant only if he is going to be late. The implication is that these premises are not mutually exclusive, and so it would be inappropriate to use Bayes' theorem to deter-

[7] The theoretical foundations of Bayes' Theorem can be found in most elementary probability texts. One such is Cramer, H., *The Elements of Probability Theory,* John Wiley & Sons, New York, 1955.

[8] As one of its options, the KES shell from Software A & E uses Bayes' Theorem in an inferencing scheme called statistical pattern matching.

mine the probability that he will call. In this case, even if they were mutually exclusive, it is unlikely that we could determine prior probabilities for either of the two premises.

It is much more typical that the certainty factor associated with any given rule is a matter of somebody's best guess as to how likely something is. We saw this sort of thing in the ear muff example earlier; the certainty that I will put on ear muffs under any given set of circumstances is a matter of pure conjecture.

Further, the ways these **certainty factors** are combined may vary tremendously from domain to domain.[9] Certainly each expert system shell has its own way of doing this. Some may try to approximate a formal theory, others may simply average the certainties of all the factors. In any case, the way any given shell deals with certainty is likely to be oversimplified, and the detailed scheme for dealing with certainties probably will have to be hand-tailored for each domain. As with conflict resolution, such tailoring will have to be based on an experiential comparison of the system's performance against the experts' performance.

5.4 Levels of Knowledge

I have on several occasions referred to control strategies for the inference engine, as well as to the need for fairly detailed background knowledge in the system. These are each related to a different level of knowledge. In any knowledge-based system there may be three levels of knowledge: decision knowledge, support knowledge, and meta-knowledge. In the real world, the difference among these types of knowledge is seldom clean-cut.

5.4.1 Decision Knowledge

For the most part we have been concerned with what I would call basic **decision knowledge**—that is, the rules that are used more or less explicitly by the expert in reaching a decision. This is the only level of knowledge which is a *sine qua non,* and it is typically the only level of knowledge which can be specified in an expert system shell.

Here is yet another example of a decision rule, one that we might see in a medical expert system:

```
IF the patient is under 13 years of age
 THEN don't prescribe Tetracycline.
```

This is a perfectly adequate rule if our main or only goal is to determine what drug is to be prescribed.

[9] One of the best treatments of certainty factors is in Buchanan & Shortliffe (1984), Op. Cit.

5.4.2 Support Knowledge

If in the knowledge-based system we are concerned not only about what drug to prescribe, but also about the reasons why something should or should not be done, then we need a deeper level of knowledge which I call **support knowledge.** In the above example, the decision knowledge would not include the ability to answer the question: Why should we not prescribe tetracycline to children under 13?

Experts or others involved in developing an expert system almost always wish to include support knowledge, and sometimes it seems so important that decisions cannot be made without it. One must be careful in such situations to differentiate between knowledge which would be helpful and interesting, and knowledge which is actually needed to make decisions. The difference between the two is not always clear.

5.4.3 Meta-Knowledge—Rules About Rules

Yet another level of knowledge is what I would call **meta-knowledge,** which affects control of the decision-making process rather than the decision itself—that is, meta-rules are those rules that affect the way the inference engine uses decision rules.

Some of this knowledge may be hard-wired into the inference engine; in expert system shells this is typically the case. In a home-grown inference engine, you have complete freedom not only to hard-wire your own assumptions, but also to use IF-THEN rules at the meta-level. Meta-knowledge concerns all those issues of certainty, inference technique, and conflict resolution which we have discussed previously.

Here are two examples of of meta rules:

- In a knowledge base in which we have predetermined the lengths of certain chain paths, we might sometimes want to change the direction of chaining.

IF forward chain is the current strategy
and IF the backward chain length is less than three
THEN temporarily change the current strategy to backward chain.

- Suppose that the truth of certain premises would make certain rules more likely to be executed that others:

IF there are rules which are relevant to ear muffs
and IF there are rules which are not relevant to ear muffs
and IF the current goal is ear muffs
THEN use first any rules relevant to ear muffs.

Meta-rules add a quality of domain specificity to the inference engine even if the specifics are not hard-wired into the code. A problem with expert system shells is that they are not able to include such information for the sake of efficiency and for the sake of long-term maturity of the knowledge base. This can be a real shortcoming compared to a human's expertise, since an expert's real expertise often comes not just from the decision knowledge, but also from how that knowledge is applied.

5.5 Explanation

We would like to have many things *explained* by the expert system, including questions that would require support knowledge.

One of the most important enhancements to any inference engine is the ability to answer questions from the user about why it has asked a particular question or how it reached a particular decision. Without such a capability it becomes virtually impossible to trace the system's logical processes. The difficulty of doing this manually arises mainly from the heuristic nature of knowledge bases in general.

As we have seen from drawing DFDs for the logical content of the knowledge base, the knowledge base is essentially a program when it is implemented. As such, it is susceptible to the possibility of errors in its logic in the same way as any COBOL program is. In fact, the knowledge base must be debugged and enhanced in the same sense as a COBOL program. And we all know that in order to debug a program, we must be able to trace through the program statements as they are or would be executed on the computer.

To debug a COBOL program we can sit down at a desk with the program listing and, by carefully keeping track of how certain data are changing, find errors simply by following the program statements. If we were to attempt this same kind of debugging on an expert system with as few as a dozen heuristic rules, we could become hopelessly lost in a very short time. Even if we could desk-check a heuristic program, it would take so long to observe the effects of adding and deleting rules that the desk-checking protocol is impractical at best.

For these reasons and others, knowledge-based system technology has been born with an understanding of the need to have a certain kind of self-awareness built into the system. In most rule-based expert system shells, the kind of explanation we can expect is a very simple trace of the rules which were invoked to reach a decision. As simplistic and often cumbersome as this is, it provides crucial information we need to trace system logic.

We would like to have more, however, and some systems give more. One of the nicest approaches is to provide a graphic structure editor which works on the rule base in the form of an AND-OR tree.[10] This is also part of the approach used by the KEE[11] knowledge engineering environment. When asked a "why" question, KEE can display both the part of the tree which was used in making a decision and the specific rules involved.

Including such an explanation capability in your home-grown inference engine is an important task that must be done carefully, but it is not as complex as the inferential parts of the system.[12]

[10] Some of the internal expert system work done at General Electric in Schenectedy, N.Y. involves such an editor, which one person referred to as "the *VisiCalc* of expert systems."

[11] IntelliCorp, Menlo Park, Calif.

[12] Once again, Buchanan & Shortliffe (1984) Op.Cit. have several helpful sections on rule-based explanation facilities. If you are working with PROLOG, Clark & McCabe (1984) Op.Cit. deals very explicitly with "why" and "how" explanations in PROLOG.

5.6 Other Enhancements

As you can imagine, the enhancements we have discussed so far can substantially increase the size of the inference engine program. However, the list of other possible enhancements can be long, including some features we discussed in Chapter 5. Consistency checking of the rule base, for example, as well as rule compression and other aids to the knowledge acquisition process, are important facilities for an expert system. Also included on the list would be learning facilities, however primitive, and a flexible user interface, possibly even natural English.

Many, if not most, of the features we have discussed, including basic inference techniques, are still subjects for research in AI. The techniques I have mentioned here have been out of the lab long enough to give evidence of fairly broad applicability in the business world. Still, each of the techniques and tools must be tailored to each specific expert domain if we are to emulate the true essence of expertise in that domain.

6. Generalizing Rules

I have devoted quite a few pages to production rule knowledge bases, and clearly describing the technical details of such systems could easily fill several volumes. Such rules are not the only way to store knowledge, but expert knowledge is commonly stored in this manner. Even when a more elaborate scheme is needed, rules are a good place to start, and knowledge-based systems typically have a production rule component.

This section describes a variation on the production rule theme that has most of the benefits of rule chaining and also overcomes some of its limitations.

6.1 Knowledge as Vector

Rules are a special case of what might be called **knowledge vectors.** A knowledge vector table lists all of the factors affecting a decision down the left margin, and different combinations of possible values for the factors are listed in columns to the right of the factors. Below each column of values is listed the decision to be made when the values in that column are true. (We actually used this approach when discussing our Knowledge Base Evaluator (KBE) expert system in Chapters 3 and 4.)

6.1.1 An Example

Figure 7-6 shows a knowledge vector table in which the decision of whether or not to put on ear muffs is determined by three factors: Cloudiness, Presence of High Collar, and Snowing, each of which can have the values Yes or No. This table includes two columns of values, each representing a real-world case; an asterisk (*) indicates that the factor is

```
KNOWLEDGE VECTORS
Generalized Rules

VALUE       FACTOR
 *   *      Cloudy
 Y   *      High collar
 *   Y      Snow

 Y   N      Put on ear muffs

1. If high collar Then put on ear muffs
2. If snow Then don't put on ear muffs
```

Figure 7-6. Knowledge Vectors—Generalized Rules

irrelevant to the decision for the case in question. At the bottom of each column is the decision to be made when that column of values holds.

Each column of values with its associated decision is equivalent to a single IF-THEN rule. The leftmost column could be written as the rule:

```
IF I have a High Collar
THEN put on ear muffs.
```

Correspondingly, the other column information could be written as:

```
IF it's snowing
THEN don't put on ear muffs.
```

Looking carefully, you may see that these two rules are contradictory. The contradiction is more difficult to see in the rule representation than in the vector representation, where all possible factors are shown for each rule. In the knowledge vector we can see that in the left case High Collar is Yes, and Snow could be Yes, giving the decision Yes. In the right rule High Collar could be Yes, and Snow is Yes, giving the decision No! Clearly a contradiction.

This contradiction underscores the importance of having in your inference engine some ability to check such logical inconsistencies. Even in this two-rule knowledge base, I did it wrong.

6.1.2 Heuristic Decision Tables

By this time, your mind may be rebelling: "Hey! You just described an ordinary decision table (popular in the 1960s). What's so high-tech about decision tables?!" If this is your reaction, then we need to look more closely at knowledge vectors and how they differ from decision tables in the same way that production rule heuristics differ from COBOL-type deterministic IF-THEN rules: in terms of certainty, completeness, dynamism.

In the decision tables of antiquity, we were constrained to enter *all* values and well-defined decisions for *each combination* of values. Using knowledge vectors, on the other hand, we build only a limited amount of not-necessarily-well-defined experience into the knowledge base. We do not intentionally withhold information from the system, but we must be realistic about the amount of experience we can add to the knowledge base in a practical amount of time.

Clearly, the more experience we build into the knowledge base, the more reliable the decisions will be. However, a decision with only 12 factors, each of which has only 3 possible values, means that the **decision space,**—i.e., all possible combinations of values—is 531,441 combinations. To completely train the system by providing decisions for each of those cases would require almost 3 years if we did one decision every 30 seconds, 8 hours a day, 5 days a week; impractical, to say the least.

In the KBE system from Chapters 3 and 4, we trained the system to make decisions by giving it a handful of cases and then set it off to make decisions on its own. Regardless of whether or not it had rules specific to the situation we asked it to assess, it was still able to to make a decision. This is quite different from the rule chaining systems we have been discussing.

6.2 Reasoning by Analogy

The secret to this improved performance lies in the use of a somewhat different inference technique, sometimes called **induction,** which is akin to reasoning by analogy.[13] With this method, the system's work is not to rigorously follow a chain of prespecified rules, but rather to look for situations from its past experience which are like the situation you have asked it to decide on. Those past situations which are most like the current situation will most heavily influence its current decision.

This approach to reasoning allows us to specify the same rules we would in the production rule situation, but frees us from having to do exhaustive work to get the system to make a decision. In addition, the analogical reasoner has a sense of how big its decision space is and can guide the expert to enter those rules, or training situations, which the system needs to fill the gaps in its experience.[14]

[13] This is the method used by TIMM, and is one of the main reasons I have chosen to reference TIMM fairly extensively.

[14] A good overview of the relationship between inference by chaining and analogical inference is Kornell, Jim, "Embedded Knowledge Acquisition to Simplify Expert System Development," in *Applied Artificial Intelligence Reporter,* University of Miami, Coral Gables, Fla. Vol. 1, No. 11/12, Aug-Sept 1984, pp 28-30.

Also, because the system is not bound to existing rules, it can use existing training to generalize and thus suggest new rules to the user/expert.

7. Conclusion

Rules, trees, and vectors are the most popular ways of representing knowledge in expert systems. Similarly, chaining and analogical inferencing are the most frequently used methods for reaching intelligent decisions. There are many limitations to the rule approach, however, and we would like to have a representation technique that allows us to integrate rules, data, and even processes in a natural way.

The frame representation techniques developed in the next chapter have the potential to be such a medium.

Chapter 8
Frames: Data, Knowledge, and More
Further Adventures with PROLOG

In Chapter 6, Structured Analysis & PROLOG, we represented the Scheduler's knowledge of instructors, engagements, and so on as very simple lists of data for each entity. This was adequate for a demonstration of prototyping, but in order to build a useful paradigm for storing knowledge we must revise our basic ideas of the way information is stored. In particular we need to find techniques which more closely approximate human information storage than those now found in data base management systems (DBMS). We need to build a knowledge base management system (KBMS) such as was described in connection with the Scheduler system in Chapter 5.

We have studied some of the most popular techniques for knowledge representation including production rules, AND-OR trees, and knowledge vectors. Now we wish to extend those simple paradigms to include a richer assortment of knowledge and then, using PROLOG, develop a set of elementary access routines for a frame KBMS.

1. Frames Defined

A **frame**[1] is a data structure that can be used to represent any real or imagined stereotypi-

[1] The idea of frames was explicitly described first by Marvin Minsky in 1974: see *A Framework for Representing Knowledge*, MIT AI Laboratory, Memo No. 306, June 1974. This paper and his subsequent work with the frame concept have been very influential in the growing use of frame systems in commercial AI work today.

cal entity—such as something called human, or an instance of a human—or it can represent a situation or event. Frames can be put together in sequence to represent a changing situation (like a movie) or linked hierarchically or in network fashion.

1.1 Database Equivalents

Each frame is roughly equivalent to what we think of as a record in a traditional database. A system of frames is roughly equivalent to a database in which many types of logical records are linked by perhaps many pointers. The difference between a frame and a database record is in the kind of information we find and how it is stored and used.

For example, each frame can have its own unique set of **slots** for information. A slot is equivalent to a field name in database jargon, but it appears explicitly in the frame; database field names hardly ever appear explicitly in the record.

In a database record, we expect things to be in a fixed format, or in a free format with items appearing in a specific order; that is, the contents of the typical database are well-defined and hence there is no need for field names in the record. In a frame system, however, each frame represents a different viewpoint, often of the same situation, and the frame is responsible for maintaining a description not only of its own viewpoint, but also of its own structure. For example, two frames—each representing an individual human—may have completely different slots depending on each individual's important characteristics.

Another difference between a frame base and a database is that each slot can be associated with, and contain, many different **types** of information. There is no real database equivalent to a frame base type since most databases store only one type of information: a data value. Data values are certainly one of the types of information which can be stored in a frame, but frames can also include such things as default values, validation criteria, executable processes, or even production rules.

Finally, of course, each slot must contain the actual **value** for each type in each slot; this value is equivalent to the data values in a database record.

As you can see, a frame base has all the capability of a relational database representation without the restrictions of pre-defined formats, types, or record content. A frame base, or frame system, is actually a symbolic relational database system that is both self-defining and in many ways potentially self-maintaining.

1.2 An Example

Figure 8-1 shows a system of three frames in a hierarchical description. The topmost frame is identified as Human and has two slots that describe what we mean by Human: # Legs and Weight.

The drawing shows that the # Legs slot contains a type of information which is a hard-wired data value of 2—that is, the Human unequivocally has 2 legs. However, just because this frame has an absolute value for number of legs, frames for individual humans

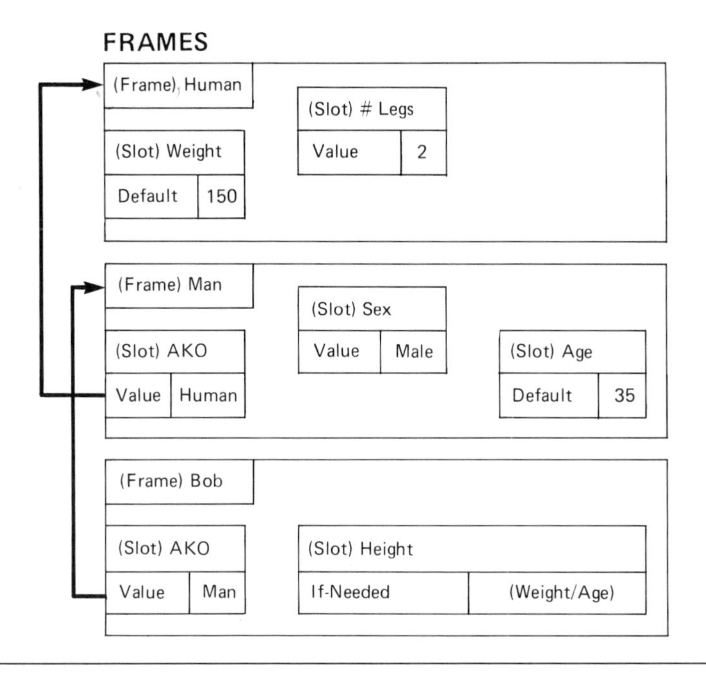

Figure 8-1. A Hierarchical Example

could have a different number; the same slot can appear in different frames and even have different meanings in different frames.

The Weight slot in the Human frame contains a **default type** of information with a value of 150. Note that this initial state of the structure and content of the frame system may change during execution, by adding or deleting slots, or by changing the values and types of information in the slots.

The Man frame also has some value and default information. It has also a slot named **AKO.** This stands for "a kind of" and is the upward hierarchical pointer in this frame system. That is, the value in the AKO slot points to a frame higher in the hierarchy of frames. Other kinds of pointers may point downward or to other frames in other hierarchies. We can imagine, for instance, that the Man frame might also appear in a hierarchy headed by an Employee frame but without a Human frame.

Finally we come to the Bob frame, a unique instance of a Man which is a unique instance of a Human. The Bob frame also has an AKO slot and, in addition, a Height slot with an **If-Needed** type of information. The value that goes with If-Needed is a procedure, which in this case is defined to be (Weight/Age). This procedure allows us to retrieve a value for Bob's Height even if one is not given as an explicit Value in the Bob frame. A further problem occurs in our example, however, since neither the Weight nor Age slots exist in the Bob frame. This system's protocol for accessing information is such that the

AKO slots will be followed automatically until Weight and Age values are obtained and then the computation will take place.

If-Needed is just one example of the kinds of processes, called **daemons,** which appear in frame systems. Others, such as **If-Removed,** might run when a value is deleted from a slot, and **If-Added** might run when a value is added to a slot. There are no limits on the types or complexity of processes that can function as daemons. Object-oriented system development environments such as Smalltalk, developed by Xerox, effectively use a frame representation scheme, and do all of their processing through daemons which are called **methods.**[2]

1.3 A Caution

Clearly, a great deal of care must be exercised when designing a frame system. They offer so much flexibility that one can imagine once again getting into a "Turing Tarpit" where everything is possible and nothing is easy. However, the problem is of the same order of magnitude as any analysis and information modeling task, for we are planning simply to replace our traditional DBMS with a frame system KBMS. We are both helped and hindered in this by the nature of symbolic information.

The word *symbolic* is used widely in AI work to refer to entities that have no intrinsic meaning, but only that which is given to them. It is this characteristic which gives us the flexibility we want in frame systems and which we used freely in defining PROLOG relations. "Human" in our example means an entity with a characteristic "# Legs" having a value of 2, and a characteristic called "Weight" with a default value of 150; it means nothing more or less than this until given further meaning or until used by a process which gives it meaning.

2. KB Access Method Requirements

In any information management system, a certain set of routines actually manipulates the physical and logical structure of the information base. I call this set of routines an **access method;** it contains such low-level detail as reading and writing physical records that we really don't want to worry about when building applications. That is, without the access method it would be very cumbersome to do any applications.

This is particularly true of our frame system KBMS where we not only have to add, delete, and retrieve information, but we first have to figure out where it is, and in accessing the frame system we may need to invoke one or more daemon processes. So,

[2] Object-oriented programming is a new way of looking at program organization and development. Among other benefits, one gets highly reusable software, generic code, and excellent prototyping capabilities. Although the object oriented approach is relevant to prototyping a DFD specification of a system, I have chosen to focus on PROLOG because of its direct connection with expert systems. Thus, a detailed investigation of object-oriented programming is outside the scope of this book.

An excellent description of Smalltalk is available from Digitalk, 5200 W. Century Blvd, Suite 250, Los Angeles, CA 90045. Their IBM PC implementation of Smalltalk is called "Methods."

while the basic functions we need are in many ways the same as in a database access method, our knowledge base access method (KBAM) needs to do more, and to do the same functions more elegantly. This is the same access method as was illustrated for the Scheduler system of Chapter 5 as Process 2.2, text Figure 5-9.

Looking specifically at the way we have defined frame system processing, there are three basic functions we need: retrieving, adding, and deleting information from a specific frame. Beyond this we need to account for the fact that defaults or daemons such as If-Needed may actually provide information when a value is not available, and finally implement the inheritance of values by way of the AKO slot during retrieval.[3]

3. Frames in PROLOG

Our previous frame example used a graphic representation which, while easy for people to use, is a bit tricky for the computer. A translation is in order. We shall use nested lists of items that PROLOG is comfortable processing.

3.1 Lists and List Processing

Learning languages like LISP and PROLOG is easy and straightforward, except that you must understand how they process lists of items and the concept of recursion, neither of which is common in traditional programming languages.[4] If you are already familiar with these concepts, you may wish to skip this section. If not, let me say that both concepts relate to things you do every day. This section simply points out the ways that computer list processing and recursion are similar to your daily activities; I shall refer to them in subsequent PROLOG examples.

Figure 8-2 shows a familiar list processing situation. We have a SHOPPING list with three items—BANK, GROCERY, and HARDWARE—which happen to be types of stores but which actually have no meaning until we need to use them as part of our processing. They serve also to point us to other lists, named BANK, GROCERY, and HARDWARE.

This is exactly the kind of list we refer to when talking about list processing in the computer. In PROLOG or LISP, the SHOPPING list might be represented by a parenthesized list consisting of its name, SHOPPING, followed by a list of its three elements—this representation is also shown on Figure 8-2.

Each of the items on the SHOPPING list is itself a list which has certain items on it. So you might say that SHOPPING is a list of lists. Using the parenthesized notation,

[3] There is no one right or standard way to do frame system manipulation. The way I have chosen here is very similar to that described by Patrick Winston in his LISP book: Winston, P.H. and Horn, B.K.P. LISP, 2nd ed. (Reading, MA: Addison-Wesley, 1984), pp. 311 ff. I have chosen to implement in PROLOG essentially the same KBAM functions that Winston has done in LISP to facilitate comparison of the two languages. There is nothing magically correct or even complete about either the LISP or PROLOG versions, but they do provide a good experimental "starter set" of working routines in either language.

[4] PL/I has explicit features to allow one to do list processing and recursion although it tends to be somewhat cumbersome. Other languages, such as FORTRAN, have no explicit list processing or recursion features, but people have written assembly language subroutines to implement these features. Nonetheless the use of these features, even in PL/I, is far from common.

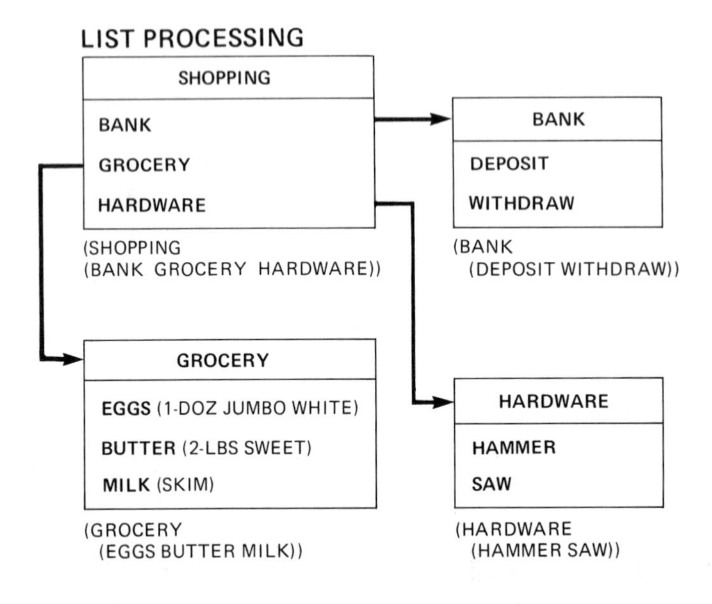

Figure 8-2. List Processing

we could show the SHOPPING list as a set of **nested lists** where the original SHOPPING items have been replaced by the lists they represent, as follows:

```
(SHOPPING
     (BANK (DEPOSIT WITHDRAW))
     (GROCERY (EGGS BUTTER MILK))
     (HARDWARE (HAMMER SAW)))
```

Some additional information on the GROCERY list tells us that each item has certain properties associated with it. EGGS, for example, have the properties of being 1-DOZ, JUMBO, and WHITE. Some list processing languages have special property lists which can be used to store such properties associated with each item; LISP is such a language. PROLOG, on the other hand, does not use property lists explicitly, but we can easily accomplish the same end by putting EGGS' properties as a list following EGGS. Now the SHOPPING list looks as follows:

```
(SHOPPING
     (BANK (DEPOSIT WITHDRAW))
     (GROCERY ((EGGS (1-DOZ JUMBO WHITE))
```

```
                    (BUTTER (2-LBS SWEET))
                    (MILK (SKIM))))
          (HARDWARE (HAMMER SAW)))
```

This **list structure** contains all the information in the diagram of Figure 8-2, and in a form which is readily understood by a list processing language. It is quite useless and meaningless, however, until we have some process which uses such a structure.

Whatever other things a list processing language can do, it typically can look at a list structure and separate the first element from the rest of the list. So let's say that when we want to refer to the first item on a list, we will use a process called FIRST-OF and give it the name of some list. It will give us back the first element of the list. Similarly, let's say we have another process called REST-OF whose job is to give us everything except the first element of the list. We need also a process, END-OF, which checks to see if there is anything in the list we have given it. If the list is empty, then END-OF has the value TRUE, otherwise it is FALSE.

As an example of the use of these processes, let's say that we have defined SHOPPING simply to be a list of its three elements, (BANK GROCERY HARDWARE). In this case, saying FIRST-OF(SHOPPING) would give us the first element of the list, BANK, in return. Similarly, REST-OF(SHOPPING) would return a list, (GROCERY HARDWARE), which is everything except the first element of the list.

3.2 List Processing and Recursion

Figure 8-3 shows the definition of a process called DO-SHOPPING, which should be familiar to anyone who uses shopping lists.

It says that to do our shopping, we use a SHOPPING list. The first thing we do is check to see if the list is empty—IF END-OF(SHOPPING)—and if so we're finished. Otherwise, we do some buying, BUY-AT, at the store which is first on the SHOPPING list. After that we do the same shopping process again, but this time using the rest of the list, i.e., everything on the SHOPPING list except the first element.

Notice that in defining the DO-SHOPPING process, we have used the name of the process we are defining. This is referred to as **recursion.** This is not quite the same as defining a word in the dictionary in terms of itself, because the processor—PROLOG for example—which allows recursive definitions actually makes a second copy of the procedure at the time of the recursive call and executes it as though it were a completely separate program with completely separate input parameters.

The remainder of Figure 8-3 shows the four iterations of the DO-SHOPPING process that would occur if SHOPPING were defined to be (BANK GROCERY HARDWARE):

1. On first executing this procedure, we are working with the full SHOPPING list. The END-OF check fails and so we do the BUY-AT process on BANK, the FIRST-OF the SHOPPING list.

2. Now we DO-SHOPPING again, this time using the REST-OF the original

```
                    RECURSION

DO-SHOPPING (SHOPPING)
  IF END-OF (SHOPPING) THEN FINISHED
  BUY-AT (FIRST-OF (SHOPPING))
  DO-SHOPPING (REST-OF (SHOPPING))

1. (BANK GROCERY HARDWARE)
      BUY AT BANK

2. (GROCERY HARDWARE)
      BUY AT GROCERY

3. (HARDWARE)
      BUY AT HARDWARE

4. ( )
      FINISHED
```

Figure 8-3. Recursion

SHOPPING list, i.e., the list (GROCERY HARDWARE). Once again, END-OF fails and we execute BUY-AT using GROCERY, the FIRST-OF the current SHOPPING list.

3. On our next recursive use of DO-SHOPPING, the REST-OF SHOPPING is the single-element (HARDWARE) list. END-OF fails on this as well and we now BUY-AT HARDWARE.

4. Using the REST-OF process on a single element list results in a list which is empty. The empty list is usually written as "()", NULL, NIL, or something similar. Thus our next and final recursive call to DO-SHOPPING is working with a NULL list. As a consequence, END-OF says yes, we have come to the end of the SHOPPING list and are now FINISHED.

3.3 Lists in PROLOG

In PROLOG, lists can be written, as above, by simply listing the explicit elements of each list. However, in processing lists in general we would like to have more abstract ways of referring to patterns of elements and sublists, such as the FIRST-OF, REST-OF, and

END-OF procedures. Instead of using explicit procedures such as FIRST-OF and REST-OF in working with lists, PROLOG uses a descriptive notation for list references. Although the specific notation varies from one PROLOG implementation to another, it generally looks something like the following:

```
(x|y)
```

where *x* refers to the first element of a list, the **head** of the list, *y* refers to the rest of the list, its **tail,** and "|" means "followed by." We could read this as "(x|y) is a list consisting of the element *x* followed by the list *y*."

So, the PROLOG statement

```
(x|y) EQ (BANK GROCERY HARDWARE)
```

would result in assigning BANK to the variable *x*, and (GROCERY HARDWARE) to the variable *y*.

There are virtually no restrictions on the list patterns that can be specified in PROLOG. Here are some examples of list patterns:

- (x y|z), a list of two variables, *x* and *y*, followed by some variable list, *z*.
- (BANK GROCERY|x), a list of two elements, BANK and GROCERY, followed by some arbitrary, possibly empty, list called *x*.
- ((x y)|z), a list whose first element is a list of two elements, *x* and *y*, followed by an arbitrary list, *z*.
- ((x|y)|z), a list whose head is a list having at least one element, *x*.

We shall use many of these powerful list patterns later when defining the relations of our KBAM.

3.4 A PROLOG Frame System Structure

In order to implement the frame system structure described above, we need to represent four things in each frame:

```
Name of the frame
      Slots
Types of information
      Values
```

Since each frame has a single identifier name and possibly many slots, one way to represent each frame is with a list. The first element of the list will be the name of the frame, and each element following the name will represent a slot description; the number of such slot descriptions is unlimited. The entire frame description will be followed by the word "frame" to let PROLOG know to file it in its dictionary under "frame." So the top-level frame structure is:

```
(Frame-Name Slot-1-Descrip. Slot-2-Descrip. . . .
            Slot-n-Descrip.) frame
```

Remember that the slot itself is essentially a field name and it can store many types of information. So we can let each slot description have a structure similar to the top-level frame structure—that is, the slot name followed by possibly many information type descriptions. The slot description might look like this:

```
(Slot-Name Type-1-Descrip. Type-2-Descrip. . . .
           Type-n-Descrip.)
```

Similarly, each type description could be represented by the type identifier, such as value or default, followed by the data for that type:

```
(Type-Name Value-1 Value-2 . . . Value-n)
```

In addition, each value can be as complex a list structure as we like.

Using this convention, the three frames of Figure 8-1 would be represented as follows:

```
(Human (Weight (Default 150)) (#-Legs (Value 2))) frame
(Man (Sex (Value Male)) (Age (Default 35)) (AKO (Value Human)))
frame
(Bob (AKO (Value Man)) (Height (If-Needed (Weight/Age)))frame
```

Describing a frame structure notation was quite easy. We simply let it follow the natural patterns of the data we wished to represent. We have defined a three-level list structure in which each level uses a similar pattern for storing its unique information. *This characteristic of using similar patterns at many levels of list structure is an important guideline to keep in mind when designing list structures.* List processing programs tend to make extensive use of recursion to push deeper and deeper into nested lists. The more similar the patterns at each level, the fewer the complexities which need to be programmed into routines which process multiple levels.

4. A PROLOG Frame System KBAM

We have now laid enough groundwork to be able to write some PROLOG relations which implement our frame structure knowledge base access method. We shall approach this first by defining relations which operate on a single frame to access, add, and delete information. Subsequently we shall deal with the issues of getting default data, invoking daemon processors, and finally inheriting values through the AKO slot.

4.1 Programming in PROLOG

Let's begin this discussion by reviewing briefly the writing of PROLOG programs which we began in Chapter 6. I shall continue to work with a PROLOG front-end called apes, but for those who may be interested, Appendix C contains standard micro-PROLOG versions of all the relations defined in this chapter.

In brief, writing a program in PROLOG means making a logical assertion such as:

```
x is-the-greater-of (x y) if y LESS x
```

Although any relation can be defined, eventually all relations must be definable in terms of PROLOG primitive functions. It is actually possible, though highly unreadable, to define all PROLOG relations in terms of the primitive unification operator, EQ, whose entire purpose is to find ways of making its left side look like its right side.

In general, executing a program involves asking PROLOG to find one or more instances for which a given assertion holds. For example:

```
find(x:x is-the-greater-of (x 3))
```

asks PROLOG, "Find all x such that x is the greater of x and 3."

This query will actually be unending, since PROLOG will try to find all possible instances of x for which the is-the-greater-of relation holds. In this case it is all positive integers greater than 3.

During execution, the IF conditions are executed in a left-to-right manner, backtracking in the same way as in backward chaining inferencing. In addition, if there are several definitions for a relation, they are used in the same *physical* order that they appear in PROLOG's dictionary. These underlying assumptions provide the procedural component of a PROLOG program, and one must be aware of PROLOG's sensitivity to such orderings. PROLOG, apart from any particular implementation, is a parallel processing language. It is only because of our current dependence on fourth-generation sequential hardware that many of these ordering constraints are relevant.

PROLOG has no looping construct such as DO-WHILE or DO-UNTIL, but it allows very flexible use of recursion as a substitute.

Although PROLOG is a descriptive language, many of its primitive relations have non-logical side effects, such as reading and writing data files.

A word about style: It is possible to write micro-PROLOG in such a way that, instead of constantly assigning values to variables, each relation is defined to be a function such that the value of the relation replaces the reference to the relation in expressions. Thus,

PROLOG would come out looking a bit more like the functional LISP language. I have chosen to use the non-functional style which, although it can sometimes be hard to read given micro-PROLOG's meaningless variable names, is more readable than the functional definitions of many of the relations we'll be discussing.

With these guidelines, we can begin to write our KBAM and describe other features of the language as they are needed.

4.2 Retrieving Information

Assuming that we have some frames to work with, we need a way to get information out of a single frame; we shall call this relation **fget.**[5]

There are actually several different kinds of things we might want to retrieve, such as data for a type in a slot in a frame, or everything about a particular type in a slot in a frame, or perhaps even the entire frame. We can use the same relation name for different kinds of retrieval, but we will need different definitions for each.

To get started, let's say we have the following frame stored already in our knowledge base.

```
(Bob (Suffix (Value II))
     (Name (Value (Edwin Robert Keller) Bob))
     (AKO (Value Man)) ) frame
```

The name of the frame is Bob, which has slots for Suffix, Name, and AKO. The Suffix and AKO slots each have one value, and the Name slot has two data in its value type. The implication might be that we want to store this person's full name as one value and some nickname as another.

4.2.1 Get a Frame

The simplest form of the fget relation would allow us to retrieve an entire frame by giving just the name of a frame:

```
find(x:x fget Bob)
```

would give us the entire frame structure listed above. We could define a relation to do this as follows:

```
x fget y
     if (y|z) frame
```

[5] If you are trying to follow a comparison with Winston's treatment of these routines in LISP, you will notice that I have tried to use roughly the same relation names as his LISP functions. In many cases, however, I have built slightly different functionality into the PROLOG relations, and a one-to-one comparison may not be entirely possible.

```
& x EQ (y|z)
```

This says simply that some value, *x*, is the fget of some *y* if there is a frame in the knowledge base which has the name *y* followed by some unknown list, *z*. Furthermore, *x* will be the same as that *y* followed by *z*.

As you can see, we have gotten a lot of mileage out of the list notation. The statement (y|z) frame is in fact a fairly complex information retrieval request. It says that I want to physically read a record from the data (knowledge) base, a non-logical side effect, which is of a particular type—frame—and parse the first element value. Then I want to compare that value with some record selection criterion, *y*, which will be given at the time of query. This powerful usage of the list structure notation is very typical of PROLOG relations.

4.2.2 Get a Slot

Now let's suppose we want to retrieve all the information for a given slot with a query such as:

```
find (x:x fget (Bob Name))
```

which would give us the following response:

```
(Name (Value (Edwin Robert Keller) Bob))
```

We can do this with another definition of fget as follows:

```
x fget (y z)
    if (x1|x2) slot-of y
     & x1 EQ z
     & x EQ (z|x2)
```

This says that x will be the fget of some frame name, y, and slot name, z, if certain conditions are true: 1) there is a slot in the y frame which is a list made up of of some element, x1, followed by a list, x2. 2) In this slot x1 is the name of the slot, z, that we're looking for. 3) The desired response, x, can be made the same as z followed by the slot information list, x2.

Actually, this definition of fget doesn't stop after a satisfactory x is found, but keeps on trying to find other slots of the same name. Having implemented this definition may have some utility, but in some cases it might make sense to force apes to stop after one right answer is found. The mechanism we use to control backtracking is a relation called /. In micro-PROLOG, / is always true and has the side effect of curtailing backtracking to its left. To use it in this or any definition of fget, it would be the last condition.

```
x fget (y z)
      if (x1|x2) slot-of y
       & x1 EQ z
       & x EQ (z|x2)
       & /
```

Notice also that we have assumed another non-primitive relation, slot-of, whose work is to find the right frame and, one at a time, give back each of its slots to be checked against the rest of our criteria. Its definition makes use of the very powerful micro-PROLOG primitive, APPEND:

```
x slot-of y
      if (y|z) frame
       & APPEND (x1 (x|x2) z)
```

We used the APPEND relation in Chapter 6 to take two lists and put them together to make a third list. The name of this relation would lead us to believe that combining lists is its primary mission. However, like most PROLOG relations, APPEND can work in both directions. That is, it not only can take two lists and put them together, but can take a known list as its third argument[6] and in its first two arguments generate all possible splittings of that list. We take advantage of this splitting feature in the slot-of relation.

The slot-of definition says that for x to be a slot-of y, then first y must be the name of some frame (y | z). And given that frame, everything is a slot-of y which can be generated by APPENDing some, possibly empty, list, x1, to a list made up of x, some slot, followed by some other arbitrary list, x2. When slot-of works on our example frame, it would in turn assign each of the slots to x. In the process, APPEND would generate the following splittings of our frame, z.

```
Split 1:
x1: ()

x:  (Suffix (Value II))

x2: ((Name (Value (Edwin Robert Keller) Bob))(AKO (Value Man)))

Split 2:

x1: (Suffix (Value II))

x:  (Name (Value (Edwin Robert Keller) Bob))

x2: (AKO (Value Man))
```

6 The term argument refers to a value or variable that is made available to a relation like APPEND when it is invoked.

Split 3:

```
x1: ((Suffix (Value II)) (Name (Value (Edwin Robert Keller) Bob)))
x:  (AKO (Value Man))
x2: ()
```

As you can see, the effect is to generate each slot in sequence as a value for x.

4.2.3 Get a Type

For the next version of fget, we wish to identify a frame, a slot, and a particular type of data, and have fget return the data stored for the type. For example, I might want to:

```
find(x:x fget (Bob Name Value))
```

This definition is very similar to the previous one except that we need to check a level deeper in the list structure for type lists and use a relation called type-of.[7]

```
x fget (y1 y2 y3)
     if (z1|z2) type-of (y1 y2)
      & z1 EQ y3
      & x EQ (y3|z2)
```

In this case PROLOG looks at each type which is part of the y2 slot of the y1 frame and tests the head of that list to see if it is the type, y3, that it's looking for.

The type-of relation uses APPEND in the same way as slot-of in generating all the possible splittings of a particular slot list.

```
x type-of (y z)
     if x1 fget (y z)
      & APPEND(x3 (x|x2) x1)
```

Notice also that type-of uses the fget relation as part of its definition, but is not really a recursive call to fget. We are now on our third definition of fget, but each of them has a different number of arguments. For purposes of PROLOG's execution protocol, these fgets are treated as totally different relations which happen to share the same name. In defining type-of, we are using the two-argument version of fget, not the three-argument version of which type-of is a part.

[7] This definition is functionally the same as Winston's LISP function for fget.

4.2.4 Checking for Data

In order to be complete, we should have a four-argument version of fget in which we would specify frame, slot, type, and a specific item of data to retrieve. This seems fairly useless in terms of retrieval, but it could be used as a test to see if a particular item of data exists. So:

```
x fget (yl y2 y3 y4)
      if (z|x) fget (yl y2 y3)
       & y4 ON x
```

Here we have another essentially non-recursive use of fget to get the type list in which we expect our data item to appear. I have introduced another primitive relation called ON, which might be better named "is-a-member-of". The idea is for ON to check to see if the element on its left side is a member of the list on the right.

4.3 Creating Frame Structure

In the previous section we assumed the existence of at least one frame for purposes of retrieving data. In the real world, of course, frames will not exist until we build them. Our next set of relations, therefore, has to do with adding information into frames.

An underlying assumption here is that if there is no data for a type, then the type will not exist in the frame. Similarly, if there are no types for a slot, the slot will not appear, and also any frame will have at least one slot. For this reason, every definition of our information-adding relation, **fput,** will have four arguments: frame, slot, type, and data.[8] This leaves us with the necessity of dealing with four different situations in which to add data:

```
No frame exists.
Frame exists, but the slot we want doesn't.
Frame and slot exist, but the type doesn't.
Frame, slot, and type all exist already.
```

4.3.1 Creating a Frame

The fput relation will always have a frame, slot, type, and value access path specified. If there is no frame, then our job is very simple: take the given information and build a frame with one slot, one type, and one data item as follows:

```
x fput (yl y2 y3 y4)
```

[8] The functionality of my fput is the same as Winston's LISP function. In PROLOG, I have chosen to do the job with four separate relation definitions.

```
if /* (ADD A NEW FRAME)
 & not (yl|z) frame
 & x EQ (yl (y2 (y3 y4)))
 & (x frame) add
```

There are a few new things happening here. For one, the relation **/*** is used to enter comments. Another is the use of the logical operator **not** on the second condition, which checks to see if the frame exists. If y1 is not the name of a frame, then we continue, and in the next condition we set x to be the frame structure we want to build.

The last condition raises the issue, however, that all the list manipulation that goes on in our relation definitions has no permanent effect on list structure. We are typically just making copies of lists and rearranging them. In order to make permanent changes to our knowledge base we need a new relation, **add,** which we use to accomplish its side effect of writing its argument to the knowledge base. In this case we created the new frame structure in the variable x, and we tack on the word "frame", and tell micro-PROLOG to add it permanently.

4.3.2 Adding a New Slot

Let us look at just one more of the fput definitions, the one which creates a new slot. The other two definitions of fput are included in Appendix C along with definitions of any support relations they require.

```
x fput (yl y2 y3 y4)
      if /* (ADD A NEW SLOT)
       & not z fget (yl y2)
       & xl fget yl
       & APPEND((y2) ((y3 y4)) zl)
       & APPEND(xl (zl) x)
       & /
       & xl freplace x
```

First we check to see if the slot exists. If not, we move on and get the current version of frame y1. The two APPEND steps arrange the new slot information in the proper form and add it to the end of the existing set of slots. Finally, the **freplace** relation deletes the existing frame and adds in the one we have just created.

In this example, the control of backtracking with "/" is crucial. If we don't stop the APPENDs from generating new slots, we will wind up with an infinite number of new frames, each having an additional occurrence of the new slot.

4.4 Deleting Information

The one remaining basic function is to be able to remove data and structure from a frame. This is a bit more complex because of our assumption about having no empty frames, slots, or types. This means that in some cases we can simply remove a data item. In others we may have to delete a fair amount of list structure as well.

We shall call our deleting relation **fdel.**[9] There will be relations of one, two, three, and four arguments, depending on whether we wish to delete a frame, slot, type, or specific data item. Each of these definitions must account for the situation where there is only one or more than one of the things being deleted. If there is only one slot, for example, and we wish to delete it, then we must remember to delete the frame as well.

For example, here is the relation which deletes a slot in the circumstance where there is one slot only:

```
x fdel (y z)
      if x fget (y z)
       & xl fget y
       & xl has-length 2
       & (x frame) delete
```

Here we first check to see if the slot exists. If it does, we get the whole frame as it will eventually have to be deleted and possibly replaced. Our next check uses a **has-length** relation, which checks the length of a list, and if this frame list has only two items—the frame name and one slot—then we know the frame must be deleted.

The other versions of fdel appear in Appendix C.

4.5 Defaults and Daemons

With these very basic frame access relations behind us, we can now consider some more interesting things such as using default data if the value we seek is not present, and invoking If-Needed or other daemons.

I've made another assumption, which is that when retrieving data from a frame, our first choice is to get the defined data value if it's present. If not, we will next look for a default type of data, and finally, when all else fails, we will invoke an If-Needed daemon processor. We will control the order of these selections by using three definitions of a new fget relation called **fgetd.** If the first one succeeds, the other two are skipped.

```
x fgetd (y z)
      if x fget (y z Value)
```

[9] This is equivalent to Winston's "fremove" function.

```
x fgetd (y z)
     if x fget (y z Default)

x fgetd (y z)
     if (xl x2|x3) fget (y z If-Needed)
      & true-of (x2 x (y z))
```

The first two definitions of fgetd are very straightforward retrieval requests for value or default-type information using the previously defined fget relation. After this, however, we must not only retrieve some information, of the If-Needed type, but also execute it. This is done using the **true-of** relation.

True-of is a relation that assumes that its first argument, x2, is a relation. It assumes further that its second and third arguments, x and (y z), are the arguments of relation x2. The desirable side effect of true-of is that it actually executes the x2 relation with its arguments. This is the mechanism for invoking a daemon relation.

4.6 AKO Inheritance

As one final elaboration of the basic access relations, we need to develop a mechanism for retrieving information from the higher levels of our hierarchy if necessary. For now, we'll just show examples of retrieving values, although default data and daemon invocation are equally applicable as shown in Appendix C.

The new relation is called **fgeti** and uses the AKO slot to find its way up the hierarchy.

```
x fgeti (y z)
     if x fget (y z Value)
      & /
```

This is the same as the first definition of fgetd except that we are limiting backtracking with the / relation. This definition will hold if the value is actually in the frame and slot specified. However, a second relation is needed to climb the hierarchy using the AKO slots as stairs.

```
x fgeti (y z)
     if (xl x2) fget (y AKO Value)
      & x fgeti (x2 z)
      & /
```

This is a truly recursive definition of fgeti. The first definition is responsible for finding a value in the frame and slot it is given. If there is none, the second definition is used. The effect of the first condition is to get the AKO slot value, which is the name of the frame referenced in the AKO slot. The second condition starts the fgeti process all over again, but this time using the same slot name in the new frame specified by AKO.

4.7 What Else?

The variations on ways of linking frames and retrieving information with or without daemons are vast. The set that you may need for any application depends, of course, on the specific application. Indeed we have not even discussed the use of a sequence of frames to represent an action sequence or some other scenario.

Like nearly everything else we have discussed, the frame representation technique is a tool which you may find fits well into your plans for knowledge-based systems. Or it may not . . . yet.

5. PROLOG or LISP?

I am frequently asked whether some particular company should standardize on using PROLOG or LISP. In general I believe that it's much too early in the usage of AI to adopt standards of any kind in the sense that businesses typically use the word *standard*—that is, it becomes irrevocable. AI is changing much too fast at the basic levels of understanding and technique to adopt such a standard. On the other hand, we must make some commitments, or nothing will ever get done. So a better question is, PROLOG or LISP for what?

PROLOG seems to come as close as possible to being an ideal language, conceptually at least, for developing and working with production-rule expert systems. The entire syntax of the language is oriented around an IF-THEN style of programming, and the underlying execution protocol is the backward chaining inference, one of the most widely applicable techniques. Further, as we have seen, PROLOG can be an excellent medium for creating rapid prototypes for the DFDs of structured analysis.

Even further, several front-ends to various PROLOGs are available specifically for expert system applications, and the hardware implementations are both diverse and credible. At this time, although it is a young language and there are few standards, PROLOG appears to be here to stay.

However, compared with LISP, PROLOG is a higher-level language, which means that it makes many assumptions about the way you want things done, about the kind of inference you will use,[10] and other assumptions.

LISP is a much more general-purpose programming language, more widely used in the United States, and it has spawned an entire generation of hardware for which LISP is the equivalent of assembly language. To get a better feel for the difference in level of the two languages, one can easily imagine writing a PROLOG interpreter in LISP, while writing LISP in PROLOG would be more difficult and less desirable.

If, as is the case in many businesses today, you are interested in building highly integrated production rule expert systems with flexible user interfaces, then PROLOG

[10] It is certainly possible to make a PROLOG application do forward chaining or analogical inference, but to do this you have to fight the language's underlying execution protocol instead of letting it work for you.

may be a good choice. If, as is the case in many engineering and scientific areas, you are interested in working at the lowest possible level in inference and knowledge representation technology, then LISP may be a better choice.

Ideally, of course, a company involved with knowledge-based systems work will eventually have both of these and other technological capabilities as well.

6. Conclusion

We have seen that the frame structure representation of knowledge is a very flexible and powerful medium. In the next chapter we discuss the Knowledge-Based Information Center, which includes natural language understanding, and find that the script technology used there is very similar to the frame material we have been discussing.

Chapter 9
The Knowledge-Based Information Center
The End and The Beginning

This book has been an adventure in discovering the practical meaning and applicability of what is called "artificial intelligence," looked at from some points of view that are somewhat different from the more academic mainstream of AI literature. My orientation has been to focus on what can be done practically with AI with the tools and techniques now available—not necessarily what we may be able to do in the year 2000. In following this orientation, my hope has been to offer people who have real, day-to-day, hard-to-solve problems in their business and professional situations some idea of what they can do now with AI and why.

1. In Summary

We have seen that AI is an area of computer science that has developed over the last few decades in parallel with the more familiar data processing, although the goals of AI are somewhat more ambitious than those of DP.

We are still a long way from realizing the ultimate potential embodied in the goals of AI. However, commercially available products now allow very satisfactory natural English communication in restricted domains, as well as exceptionally cost-effective systems that perform decision support, medical diagnosis, manufacturing control, and many other professional-level tasks. Through expert system technology, it is possible to capture the experience of highly qualified professionals and use it to plan or make decisions similar to how a human expert does.

Further, our study has suggested workable guidelines for deciding whether a particular business problem is a good candidate for knowledge-based system treatment. We have also worked through examples that strongly suggest that significant benefits can be obtained from this new technology by integrating it with more traditional approaches to system development. We don't have to throw out the old in order to benefit from the new.

We are not finished, however—we have not really provided a context for the production use of knowledge-based systems or seriously addressed the subject of communicating with the computer in natural, conversational English. This chapter takes care of this unfinished business.

2. The Information Center

The information center (IC) is a central facility within a company that supports a wide variety of tools and applications, and through which end-users can directly access those facilities. Its purpose is to become a resource that people with minimal training can use to access any function the computer is capable of performing—whether it's information retrieval, problem solving, system building, or some other task. However, like the fourth-generation systems that form the backbone of most ICs, today's information centers are quite primitive compared to what they can become in the future.[1]

2.1 Toward a Knowledge-Based IC

Typically today's IC, whether centralized or distributed, consists of a handful of not-very-user-friendly, poorly integrated data base systems for report writing, decision support, graphics, statistical analysis, and similar functions. An IC may sport a wide variety of mainframe terminals and PCs. Although training is available for most of the individual IC tools, end-users are generally on their own, effectively doing their own system development with the available tools. The IC support staff, which typically numbers about one for every sixty end-users, is available to respond to questions, but has no proactive role in satisfying user needs.

These first-generation information centers are a bold and useful step beyond writing a COBOL system for every need, but they fall far short of the users' ultimate need: an intelligent information resource, or what I like to call the **Knowledge-Based Information Center (KBIC).** We must ask, therefore, how we get from the present IC paradigm to one in which the end-user needs literally no help in order to access and manipulate information of all kinds with the computer. Somehow, with all the promise of AI technology, AI should hold the key.

[1] There is a very broad spectrum of IC implementations. Cited here is an "average" over many cases I have seen. Some ICs are much more sophisticated than this, some much less.

But where is the serious growing edge of the AI utility, the applications that everyone in the office and in the information center can use? And how will it impact the end-user community and the management of data processing? Following are some guidelines for the kinds of knowledge-based tools and applications that may form the foundation of the KBIC.

2.2 Is It Expert System Products?

Guideline: The KBIC should support expert system products, and home-grown expert systems, that are truly expert and have a broad base of end-users.

As we have seen, AI technology is producing a class of computer system whose competence in specialized decision and planning tasks can rival the performance of intelligent human experts. This is of course its major advantage, but it is also a limitation which suggests that supporting a wide variety of expert systems may not be the best role for the KBIC support staff.

The main reason for this is that expertise is typically not only narrow in scope, but also narrow in applicability. For example, very few people in any company would use an expert system that helps identify interest swap partners. In fact, it's likely that fewer than 1% of the readers of this book even understand, or care to understand, what such an expert system would do. Therefore we seek expert system applications that can solve unique problems expertly. The condition of wide applicability must unfortunately take second place.

There certainly are notable exceptions—in domains such as financial planning, investment management, and many others—which will find wide acceptance as knowledge-based products, but these are indeed exceptional domains. Those knowledge-based products now coming on the market that do have true expertise as well as broad applicability are the ones that probably should be supported in the Knowledge-Based IC.

Not every nook and cranny of expertise is going to be filled by an expert system supported in the KBIC. Trying to do so seems to me neither like the growing edge of the AI utility, nor a good use of IC personnel's energies.

A closely related issue is the question of whether a company should build its own expert systems or wait for a knowledge-based product to come on the market. Naturally there is no easy answer to that question, but part of it depends on how long one might have to wait for a vendor to market a knowledge base in any particular domain of expertise.

The fact is that a vendor may never come along and build an expert system on spec for a particular domain. Although building a particular expert system may be cost-justified for a particular company, the potential market for experts in most domains is nowhere near a commercially interesting size.

2.3 How About Shells?

Guideline: Support a variety of tools that are oriented toward being expert system shells.

Another option, of course, is to include one or more expert system shells for whomever in the KBIC may fancy being a knowledge engineer. Although I say that with tongue in cheek, this is actually not a bad idea, if for no other reason than that, by quickly failing to achieve expert results from a casual effort, folks may begin to appreciate the depth of involvement required to develop a successful expert system.

On a more positive, justifiable basis, having a variety of shells available gives serious expert system developers some important alternatives for prototyping their early knowledge base work. Keep in mind, however, that many commercial shells fall into exactly the same production rule paradigm and differ very little in terms of knowledge acquisition or decision making. Be careful not to waste resources simply by buying many small variations of the same tool.

We have discussed the capabilities of several shells, which you may wish to review before acquiring any shell systems. Additionally, I suggest the following categories of shell products as being different enough from one another that an IC could easily justify supporting one of each:

- PROLOG
- A production rule chaining system.
- An analogical, or induction, reasoning system.
- A generate-and-test system.
- A Bayes' theorem, or statistical pattern-matching, system.

2.4 Knowledge Engineering Environments?

Guideline: Support at least one sophisticated software and hardware knowledge engineering environment, but not in the KBIC.

Much of this book has addressed the development of knowledge-based systems without benefit of AI workstations or low-level programming in LISP. This emphasis reflects my belief that, in the near term, integrating the new technology with more traditional environments can yield tremendous benefits.

Nonetheless, over a period of years, most companies are likely to find a need to go to another level of AI system development. In such a situation, the AI workstation hardware and software will be indispensable as an adjunct to the analysis, design, and development of systems.

However, even as this begins to happen, it is likely that the workstations themselves will not be the medium for accessing KBIC applications. It is more likely that these will be either system development tools exclusively, or that they will be used as **knowledge peripherals** by a mainframe computer.

2.5 Natural Language Interfaces?

Guideline: Every KBIC facility should be optionally available through a natural language interface.

Knowledge-based technology offers not only back-end functionality in the KBIC; it also holds the potential for allowing users to access KBIC functions through natural English. This can bring to the KBIC the most powerful kind of tool: one that facilitates the use of all the other tools. Natural English can truly bring the KBIC to a realization of its full potential, making direct computer access a possibility for the largest and most diverse group of people: the non-data processing end-user.

The choice of application, KBIC help facilities, and all other central functions should be available through natural English. It is as though the computer's operating system were turned over to end-users through a natural English interface; the KBIC is thus run by an **end-user operating system.**[2]

In the next sections, we look at two generations of natural English technology: natural language data base query (NLQ) systems, and knowledge-based natural language (KBNL) systems.

3. Natural Language Query

NLQ, the first generation of natural English system, is a type of expert system whose expertise is embodied in its rules for understanding English queries. Its main work is to act as a front-end to many other types of systems, and possibly to orchestrate the integration of those systems in response to a natural English query from an end-user. This sounds very much like our description of the end-user operating system, and in fact NLQ products can form the basis for such a facility.

3.1 A General-Purpose Front-End

An NLQ system could theoretically front-end and integrate any other processors, including expert systems, but NLQ products are typically being used for ad hoc information retrieval from the same kinds of data bases used by the so-called fourth-generation languages (4GLs), such as FOCUS, RAMIS, or EASYTRIEVE.[3] Figure 9-1 illustrates this relationship and gives a few examples of the kinds of query NLQ products can handle.

In this role, typically the only part of the NLQ system's work which is knowledge-based is the interpretation of the query, which is then translated into formal commands for one or more back-end processors. That is, once the English query has been interpreted, the system's main work is to retrieve the required records from a well-defined data base, do some often complex data processing on those records, and produce a report.

[2] Thanks are due to Larry Harris, president of Artificial Intelligence Corp., for observing the need for an end-user operating system in the IC. Artificial Intelligence Corp. develops and markets a line of natural English products called INTELLECT.

[3] The INTELLECT NLQ product actually front-ends FOCUS. Similarly, MPG ENGLISH is a front-end to RAMIS.

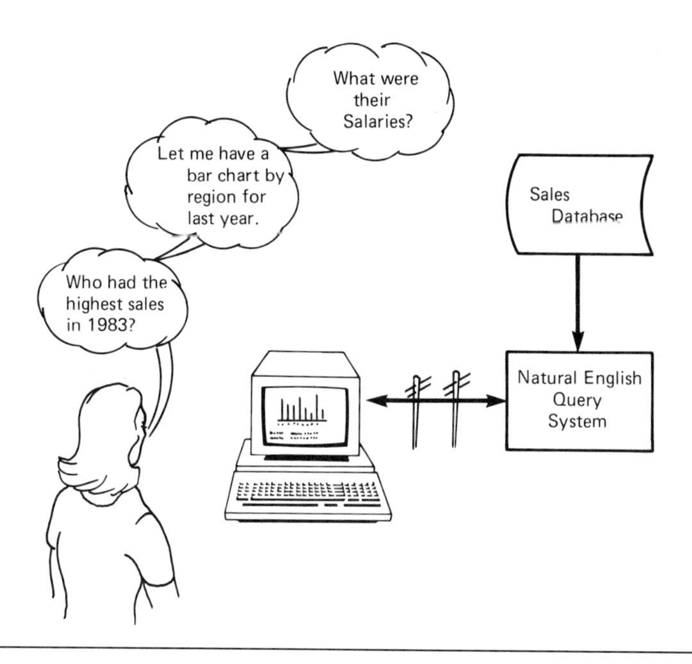

Figure 9-1. The NLQ Environment

Thus the main advantage of NLQ products—and it is a significant advantage—is to allow access to various data base files without the user having to learn a formal query language such as a 4GL.

3.2 Advantages Over Fourth-Generation Languages

Natural English access to the computer is an important breakthrough because, for the first time, end-users with little or no training can interact directly with the machine. Until the advent of NLQ products, the best access available was through fourth-generation formal query languages, which require several days of intense training followed by regular usage in order to be used effectively.

Further, using a fourth-generation language effectively requires a basic understanding of computer programming as well as an appreciation of data base structure and navigation. Figure 9-2 shows a natural English query as it would be input to an NLQ product and the equivalent 4GL code. As you can see, the natural English version makes it totally transparent to the end-user that answering the query in a 4GL requires a two-pass process on the data and the creation of a temporary file.

Another example in Figure 9-3, shows even more clearly the processing complexities from which an NLQ system can shield the end-user. In this case, the 4GL statement

NLQ Input

```
Give me a ranked percent of total
amount broken down by product type.
```

4GL Input

```
TABLE FILE PROD
SUM PCT.AMOUNT AND HOLD
BY PROD TYPE
END
TABLE FILE HOLD
PRINT PROD-TYPE
BY HIGHEST AMOUNT
END
```

Figure 9-2. NLQ vs. 4GL.

of the NLQ input includes such requirements as remembering the data base schema names, knowing that four logical files are actually involved in answering the question, and telling the 4GL how to navigate among those files to produce an answer.

The subset of English understood by the NLQ systems is fairly large, allowing questions to be phrased in many different ways. They can handle pronoun reference to previous queries and are comfortable with queries that are sentence fragments. In an IC situation, such NLQ systems can be very useful wherever an end-user needs to ask ad hoc queries of a well-defined data base.

3.3 The End-User Operating System

One of the greatest advantages of the NLQ system is its role as a system integrator.

Let's say, for example, that I want to see line graphs of exponentially smoothed sales in the Eastern and Southern regions for the last two years by month and projections for the next year. Ideally, I should be able to ask for it using exactly those or similar words. In fact, some of today's NLQ systems approach this level of integration.

Today's NLQ systems, such as INTELLECT, THEMIS, or MPG ENGLISH[4] are a

[4] INTELLECT is developed and marketed by Artificial Intelligence Corp. of Waltham, Mass., THEMIS is developed and marketed by Frey Associates of Amherst, N.H., and MPG ENGLISH is developed and marketed by Mathematica Products Group of Princeton, N.J.

NLQ

```
Show me 1980 and 1981 sales and 82
forecast in each region with names and
phones of the district managers.
```

4GL

```
SELECT OFFICE, REGION, NAME, EMPLOYEE,
       PHONE, SALES, VALUE, FORECAST
FROM OFFICE, EMPLOYEE, FORECAST, SALES
WHERE (SALES.YEAR=1981 OR SALES YEAR=
1980 OR FORECAST.YEAR=1982) AND
SALES.OFFICE=OFFICE AND FORECAST.OFFICE
=OFFICE AND
OFFICE.MGR=EMPLOYEE.EMP NUM
```

Figure 9-3. Verbosity

first-generation approximation of the end-user operating system. Their task is to accept a natural English query from the user and translate it into a series of commands that orchestrate the interplay of whatever system functions are needed to respond.

For the example above, the end-user operating system must deduce the following tasks from the English request:

- Select data base records, possibly from several related data files, in which the region field is East or South.
- Retrieve sales fields for each month of the last two years and each month of the next year's projection.
- Invoke a process to summarize those records by region.
- Format these data and pass them to an exponential smoothing process.
- Take the results of the smoothing process, format them, and pass them to the graphics processor.
- Accept the display information from the graphics processor and return it to the user as requested.

All of this orchestration, including data base navigation, is totally invisible to the end-user, whose job is simply to conceive the request and enter an English statement of it.

3.4 Limitations

In spite of its advantages over 4GL systems, one does not have to use an NLQ system very long before getting the feeling that it doesn't really "understand" in the same way that a human assistant would understand if asked the same question. This is in fact true. NLQ systems are very limited in what they can understand partly because their ultimate goal is to translate a user query into a form acceptable to a traditional data processing back-end system.

In front-ending 4GL retrieval tasks, for example, this means locating fields in multiple logical files, applying (possibly quite sophisticated) record selection criteria, invoking processing routines, and invoking display processors. The system is working with data, not knowledge, and so its understanding will be limited to what can be done with the data.

Although NLQ systems typically are programmed to avoid giving wrong answers, this limited "understanding" often prevents them from giving more than a barely adequate correct answer, which can be irritating. It is irritating mainly because end-users expect any system that claims to understand natural English to understand not just the words, but also the *intention* behind the words. Unfortunately, NLQ systems have virtually no ability to understand a user's real intention in asking a question.

For example, let's say I have an NLQ front-end to the Official Airlines Guide Electronic Edition,[5] and I want to get from Boston to Miami.

I might readily ask the question:[6]

```
DOES DELTA HAVE A NIGHT FLIGHT TO MIAMI?
```

If there is no such flight, the NLQ system will respond:

```
NO
```

This is certainly an adequate and literally true answer. However, it's not terribly useful, since I apparently have to ask about every possible airline to get the information I really want. The point is that the NLQ system has no way of knowing that what I really mean is: Are there any night flights to Miami from Boston, and if so, then I would prefer to travel on DELTA.

Not being able to understand a user's intention in asking a question is a very serious failing when it comes to understanding natural English in general, and it's the reason why, in the long run, this NLQ technology will become an historical anecdote. In the meantime, however, if the users understand this unfortunate limitation in "understanding," it may not be so bad. My experience working with NLQ systems tells me that, in spite of this

[5] The OAG Electronic Edition is a computer service that allows dial-up access to commercial airline schedule and fare information. The query language is very primitive and cumbersome, so one can imagine wanting an NLQ front-end for it.

[6] I am indebted to Ken Wirt, vice president of marketing at Cognitive Systems, Inc. of New Haven, Conn., for this example. Cognitive Systems is one of the most active and successful companies providing the Knowledge-Based Natural Language (KBNL) systems discussed in the next section.

problem, a large percentage of the queries most people wish to ask of a data base can be handled by an NLQ system.

3.5 How Do We Get NLQ Capability?

There are two ways to get NLQ capability: one is to buy it; the other is to build it.

At this time, if you want a general NLQ capability for your information center, I would strongly recommend buying it. INTELLECT, for example,[7] required hundreds of work-years to develop it to a point where truly uninitiated end-users can use it reliably. To try to build this same level of performance into an in-house system would not be cost-effective.

On the other hand, if you want natural English capability for an expert system you are building, then building the natural English piece may make sense. This is because simple NLQ approaches to understanding restricted domains of natural English are well-understood and can be made available almost anywhere you want NLQ. It can be read about and studied from many sources.[8] Thus building a domain-specific English front-end for one system may be more cost-effective than paying an NLQ vendor to try to integrate your application requirements into its product.

4. Knowledge-Based Natural Language

Now let's suppose that we have an application—managing stock portfolios and watching the stock market for example—in which it's important to us to have the investor communicate in natural English, but the NLQ limitations are just too severe. We need the system to understand the context of the stock market in the same way the investor does.

4.1 A Stock Market Advisor

For example,[9] let's suppose that the system has assessed investor preferences and circumstances, as determined from an earlier natural English dialogue, assessed current market circumstances, and created a "buy list" recommendation, as shown in Figure 9-4. The investor, however, suddenly remembers Uncle Charlie's advice from long ago that you

[7] I often use INTELLECT for examples of NLQ systems partly because it has set the standard for such products, and partly because of my familiarity with the product resulting from my several years as an executive at Artificial Intelligence Corp.

[8] If you are working in LISP, there are many public domain natural language possibilities; contacting universities such as MIT or Stanford may be a good approach. Also the *LISP* book by Winston (1984) Op. Cit. contains some good text on natural language understanding, as does another excellent LISP tutorial: Charniak, E., Riesbeck, C.K., and McDermott, D.V., *Artificial Intelligence Programming,* Lawrence Erlbaum Associates, Hillsdale, N.J., 1980. If you're working in PROLOG, you're using a good medium for natural language work—much of the early use of PROLOG has been in this area. Of particular interest may be the previously referenced micro-PROLOG work by Clark & McCabe (1984) Op.Cit. and the *PROLOG-86 User Guide & Reference Manual,* Solution Systems, Norwell, Mass. 1984, which includes a small natural language parser in PROLOG-86.

[9] These examples are from actual sessions with LE COURTIER, a 1984 Cognitive Systems advanced prototype stock portfolio advisory system.

should never trust a utility,[10] and rejects the recommendation, saying: "No, I don't like utilities" or "Forget utilities," or something like that. Our KBNL system understands the user's view of the situation and responds by accepting the user's input *in the sense that it was intended: as an additional heuristic to be applied when selecting stocks.* It therefore creates the modified recommendation shown in Figure 9-5.

The important comparison here is between the KBNL response we have just seen and the likely NLQ response to the same statement. An NLQ system would treat the user's statement, "I don't like utilities," either as meaningless input or as a request to list everything about utilities stocks. Clearly, these are both silly interpretations, given the user's understanding of the portfolio context.

Figure 9-6 shows two other examples of KBNL responses to questions which require a knowledge-based understanding of the context—the domain of expertise—in which the conversation is taking place.

To the first question—"Should I buy SGB?"—the NLQ system, assuming it has a data field which can be retrieved to indicate the desirability of a stock, would respond either YES or NO. As you can see, the KBNL responds that SGB is not a bad investment, and then it tries to be more helpful—since the user has asked about a stock in the Banking sector, it makes an alternative recommendation in that same sector.

The second example—"What is the P/E for SGB for the last 5 years?"—points out another benefit of the KBNL approach to query situations—it has a good idea of what it does not know. For this situation, the NLQ system would probably answer that NO RECORDS MEET SPECIFICATIONS if it has no history of P/E. The KBNL system once again takes the approach of being helpful in context, acknowledges what it does not know, and offers some related information which could be helpful.

4.2 Knowledge vs. Data

As you can see, the KBNL system can answer all the same kinds of questions as the NLQ system, but it goes a step further. It includes extensive knowledge of a particular domain, including the reasons why people would ask certain questions, and how to be helpful in areas of uncertainty.

For another example, let's go back to the OAG Electronic Edition. We said that giving NO as an answer to the question DOES DELTA HAVE A NIGHT FLIGHT TO MIAMI? was less than satisfactory. Would it be more satisfactory to give the answer:

```
NO, BUT DELTA HAS A NIGHT FLIGHT TO SAN FRANCISCO.
or NO, BUT DELTA HAS A NIGHT FLIGHT TO MIAMI FROM SEATTLE.
```

Of course not. The point is that even having the system search out other possible data-oriented answers to the question is likely to turn up answers which are off the mark from what the user wants. In order to successfully and adequately answer natural English

[10] Once again I am indebted to Ken Wirt of Cognitive Systems for this example.

I would recommend making the following stock purchases:

Sector	Company	# shares	Current Value	New Value
Oil	Sidro	buy 36		49,680
	Petrofina	buy 9		45,360
Banking	Kredietbank	buy 9		51,030
Utilities	Unerg	buy 37		49,432
Insurance	A. G. Fin.	keep 50	380,000	380,000
Const. Materials	C.B.R.	buy 26		48,620
	Current portfolio value		380,000	
	Proposed portfolio value			624,122

****

Are you satisfied with this recommendation?

==> *no, i don't like utilities*

I am now taking into consideration the preferences you have given me as I decide on my advice.

Figure 9-4. Le Courtier Recommendation

I would recommend making the following stock purchases.

Sector	Company	# shares	Current Value	New Value
Oil	Sidro	buy 43		59,340
	Petrofina	buy 11		55,440
Banking	Kredietbank	buy 14		79,380
Insurance	A.G. Fin.	keep 50	380,000	380,000
Const. Materials	C.B.R.	buy 26		48,620
			-------	-------
Current portfolio value			380,000	
Proposed portfolio value				622,780

****

Figure 9-5. Le Courtier Modified Recommendation

```
Please ask your questions here. When you are
done, press F1/RESUME. Remember, you can also
initiate a question session by pressing F2/
QUESTION.

  **************************************

==> SHOULD I BUY SGB


Soc. Gen. de Banque would not be a bad invest-
ment (it has average market performance); how-
ever, Kredietbank would make a better invest-
ment in the Banking sector.

Press F1/RESUME when you are done asking ques-
tions.
  **************************************

==> what is the p/e for sgb for the last 5 years


I have no historical information yet. However,
this current data may be useful:

The price-to-earnings ratio for Soc. Gen. de
Banque is 6.30.


Press F1/RESUME when you are done asking ques-
tions.
  **************************************
```

Figure 9-6. KBNL Helpful Responses

questions in a domain, the system must have an expert's understanding of *everything* that makes sense in that domain.

4.3 The Need for a Script

In this case, and in many query situations, there is not enough information either in the question or in the data base to allow an in-depth understanding of the question. Yet in spite of this we desperately want to be able to give the user a meaningful answer.

The KBNL technology does this by keeping track, in a knowledge base, of the many scenarios which can possibly sensibly happen in a particular domain. These data are not in the data base—they are possible scenarios for a given domain called **scripts.** And it is through very extensive scripts for each domain that a KBNL system is able to understand the user's intention in asking a question, or to take actions implied by a particular statement.

In short, real understanding means having a thorough knowledge of a particular context. Almost every situation we encounter has more-or-less well-defined scripts. In fact, as humans we are able to get along in the world because we understand the scripts that underlie the situations in which we find ourselves. Expertise in a given situation often depends on how well one understands that situation's unspoken behavioral assumptions: its script.

In the OAG example, the problem is fairly simple: the system simply needs to know that if a question has been asked in which a destination city have been mentioned, then it's likely that the user's intention is to discover facts about getting to that destination from some home base and to rule out answers which don't serve that end.

4.4 How Like an Expert System

The knowledge underlying expert systems is quite similar in its nature and complexity to KBNL script knowledge, although the goals of the expert system are somewhat more narrow.

The main purpose of an expert system is to perform complex decision and planning tasks in the same way a human expert does; when its knowledge base matures to a truly expert status, one would expect to have the option of replacing the human expert. KBNL systems are concerned also with knowledge-based decision making and planning. However, they explicitly add to this the ability to communicate in comprehensive natural English and are more likely to be used as an intelligent apprentice or advisor.

4.5 A KBNL - NLQ Comparison

We have seen that NLQ systems are an important advance in user-friendliness and system integration for the KBIC. In terms of functionality, however, they offer very little that isn't also available through the currently popular English-like 4GL languages. As we have observed, they are also severely limited in their ability to "understand" the user because the only data they have available are the facts stored in a data base.

To illustrate the need to overcome this limitation, consider the following questions that we might wish to ask of an inventory management system questions.

HOW MANY IC96's DO I HAVE IN STOCK?

This question is easy for an NLQ system, since stock levels probably appear directly in the data base. But how about:

IS THAT ENOUGH TO MAKE 300 DOZEN LEVEL 2 TRACTORS?

This question is somewhat more difficult for an NLQ system, since it requires not only stock data, but also an understanding of how to use a bill of materials for a particular manufactured item. Typically this level of knowledge is beyond the abilities of current NLQ systems.

EASYTALK[11] is an example of a KBNL system. The KBNL system has access to the same data as an NLQ system, but in addition it has a comprehensive knowledge of conventional relationships among data in the subject area. This allows it to deal with questions that require information not explicitly contained in the data base.

For example, in a KBL accounting package, I could ask not only about manufacturing requirements of level 2 tractors, but also questions which require the system to have an understanding of my stocking philosophy, expected part failure rates, and the like. For example:

HOW MANY OF THESE [IC96's] SHOULD I STOCK TO BE SURE
THAT I WILL BE ABLE TO PROCESS AN ORDER OF THIS SIZE NEXT TIME?

Or, consider the question:

HAVE ANY OF MY RATIOS CHANGED SIGNIFICANTLY SINCE LAST MONTH?

Although some of the foregoing questions could possibly be handled by an NLQ if the data base had been contrived properly, this last question is completely outside the scope of NLQ capabilities. The reason lies in the word SIGNIFICANTLY. In order to understand that word correctly, the system has to realize that it means something different for each accounting ratio and be able to apply the correct interpretation to each ratio.

5. The Impact of AI on DP Management

AI in all of its forms is guaranteed to bring significant changes into every data processing organization. But what's new? Every technology advance seems to louse up everybody's

[11] EASYTALK is part of a comprehensive accounting package developed and marketed by Intelligent Business Systems of New Haven, Conn.

carefully thought out five-year plan. It's just that now, it's lousing up not only the five-year plan, but the one-year plan as well. What's to be done?

It is a given these days that there is absolutely no way to keep up with technology. Five years ago, major technological changes were happening every 2–3 years, now it's down to every 6–12 months. Not only is it impossible to keep up with that kind of dynamic environment, but it's absurd to try. If you're worried about the competition, don't. They have the same problems that you have.

Whatever expert system shell you buy will probably be obsolete within a short period of time, but this should not be a worry because obsolescence is very often in the eye of the beholder. You can do wonderful AI applications in BASIC or FORTRAN or PL/I using IBM 4341s or PCs or whatever you have available. When you can afford to change from what you are using now, you will probably not be the last one to make that change, and you will benefit significantly by changing when it makes sense for you to do so. If, however, you try to change with every change in technology and hence change too often, or before you are ready, you will simply create chaos in your shop and lose in the long run.

For years DP management has been overwhelmed with special requests for reports, data base updates, and system maintenance tasks for end-users from all over the company. The backlog of such requests in some cases exceeds three years of elapsed time. Even worse, in addition to the relatively straightforward reporting and file management requests, company management is desperately crying out for automated solutions to decision support problems that can only be handled by advanced technologies—technologies for which there is no time or budget for study and implementation. And in the personal computer realm, data processing is supposed not only to assess, make available, and support hardware from many different manufacturers, but also support user-developed applications without having been involved in or even informed of their existence. In some cases, the problem is crippling.

Much of the new technology can help with this set of problems. But every new technology brings the responsibility for creatively managing its integration into existing DP operations, and AI technology is no exception. As one would expect, there is the question of whether to buy or build; for AI today, the answer is almost always to both buy and build. As a consequence, DP management needs to become aware both of commercial AI offerings and of the resources needed to do AI system work in-house.

My advice is to continue to focus on solving business problems in the best way you can with what's at hand, but stay aware of what's coming along technologically and evaluate it carefully for applicability in your circumstances. By managing change well, based on the experience gained over the years, you will continue to succeed beyond those who can't wait to try something new and popular which is the wave of the future.

6. What's Next?

The technology of AI has (finally) arrived and, although still in its infancy, is already producing large payoffs for those companies that have pioneered its commercial application. Companies now beginning to implement AI programs can develop a significant competitive edge over the next five to ten years through improvements in costs, revenues, and productivity, or by being able to solve problems that were previously too complex.

At best, this is the time your company should be starting something with AI technology. It's not necessary to undertake a large knowledge-based system to gain an appreciation of the technology's potential, or to reap some of its benefits. It is important at least to have some person in your organization who has some orientation to the subject and who can maintain a current reading acquaintance with the field.

The new age of computing is upon us. It is past time when any company can shrug off AI/Expert Systems as something to look into tomorrow; tomorrow has become today. As always, however, the past is prologue, and must be the foundation on which the future is built. Look on this new technology as a set of tools and techniques; based on experience choose the ones that fit and that can improve today's systems, and discard the rest.

APPENDIX A
KBE DECISION STRUCTURE

In this appendix I have listed the structure of each of the mini-systems used for KBE in chapters 3 and 4, as illustrated on Figure 3-4.

In each case I have listed the DECISION, which is the dataflow output from one of the mini-systems such as WORTH, along with the values that dataflow can have. Following the DECISION is listed each of the FACTORS which affect the DECISION and the allowable values for that factor.

This type of explicit structure listing is unique to TIMM, although many other expert system shells have textual or graphic means for displaying an equivalent structure. Some show an AND-OR tree structure to represent the logic of the rules.[1]

```
**********************************************************************************
                                  Corporation
                    TIMM (TM) The Intelligent Machine Model
                                  Version 2.0
**********************************************************************************


   DECISION STRUCTURE FOR THE EXPERT SYSTEM "DETERMINE WORTH"


DECISION:
        WORTH
           Choices:
            NEGATIVE
            LOW
            MODERATE
            HIGH

FACTORS:
        PAYOFF/COST
          Values:
            UNDER 1
            1-1.5
            1.5-3
            OVER 3
```

[1] AND-OR trees are discussed in Chapter 7, Inference & Knowledge.

```
PERCENT SOLUTION
  Values:
    UNDER 50%
    50%-75%
    75%-90%
    OVER 90%

TYPE
  Values:
    DEMO
    USEFUL
    NECESSARY
    CRUCIAL
```

DECISION STRUCTURE FOR THE EXPERT SYSTEM "DETERMINE COMPLEXITY"

```
DECISION:
        COMPLEXITY
           Choices:
            LOW
            MODERATE
            HIGH

FACTORS:
        INTUITION/COMMON SENSE
          Values:
            UNDER 10%
            10%-50%
            OVER 50%

        TECHNOLOGY
          Values:
            BUILD
            ENHANCE
            MODIFY
            EXISTS

        DECISION DEFINITION
          Values:
            FUZZY
            OK
            WELL-DEFINED

        KNOWLEDGE DOMAIN
          Values:
            ECLECTIC
            OK
            NARROW
```

```
DECISION STRUCTURE FOR THE EXPERT SYSTEM "DETERMINE EXPERTISE"

DECISION:
        EXPERTISE
           Choices:
            UNAVAILABLE
            OK
            AVAILABLE

FACTORS:
        NO. OF EXPERTS
          Values:
            0
            1-3
            OVER 3

        EXPERT AVAILABILITY
          Values:
            UNDER 50%
            50%-75%
            75%-90%
            OVER 90%

        TURNOVER
          Values:
            LOW
            MODERATE
            HIGH

        EXPERT ATTITUDE
          Values:
            HOSTILE
            INCOMPETENT
            UNINTERESTED
            WILLING
```

DECISION STRUCTURE FOR THE EXPERT SYSTEM "ASSESS CONTROLS"

```
DECISION:
        CONTROL
           Choices:
            LOOSE
            OK
            TIGHT

FACTORS:
        DATA CONTROL
          Values:
            LOW
            MODERATE
            HIGH

        PROCEDURE CONTROL
          Values:
            LOW
            MODERATE
            HIGH

        PERFORMANCE METRIC
          Values:
            LOW
            MODERATE
            HIGH
```

DECISION STRUCTURE FOR THE EXPERT SYSTEM "ASSESS RISK"

```
DECISION:
        RISK
           Choices:
            LOW
            MODERATE
            HIGH

FACTORS:
        COMPLEXITY
          Values:
            LOW
            MODERATE
            HIGH

        EXPERTISE
          Values:
            UNAVAILABLE
            OK
            AVAILABLE

        CONTROL
          Values:
            LOOSE
            OK
            TIGHT
```

```
*******************************************************************************************
                    (c) Copyright 1984 General Research Corporation
                     TIMM (TM) The Intelligent Machine Model
                                   Version 2.0
*******************************************************************************************
```

DECISION STRUCTURE FOR THE EXPERT SYSTEM "DETERMINE SUITABILITY"

```
DECISION:
        SUITABILITY
           Choices:
            POOR
            FAIR
            GOOD

FACTORS:
        WORTH
          Values:
            NEGATIVE
            LOW
            MODERATE
            HIGH

        EMPLOYEE ACCEPTANCE
          Values:
            NEGATIVE
            NEUTRAL
            POSITIVE

        SOLUTION AVAILABLE
          Values:
            ADEQUATE
            PARTIAL
            NONE

        EASIER SOLUTION
          Values:
            NONE
            PARTIAL
            COMPLETE

        TEACHABILITY
          Values:
            DIFFICULT
            POSSIBLE
            FREQUENT
```

```
RISK
  Values:
    LOW
    MODERATE
    HIGH
```

APPENDIX B
KBE RULE BASE

This appendix lists for each mini-expert system of Figure 3-4 the IF-THEN rules which are the mini-spec for that mini-system. In the listing, each rule has listed all of the factors which may affect the output dataflow decision. Some of the factors, such as the PAYOFF/COST factor in Rule 1 for the Determine-Worth mini-system, have actual values associated with them: OVER 3 in this case. Referring once again to Rule 1 for Determine-Worth, the other two factors, PERCENT SOLUTION and TYPE, have an asterisk, *, as their value. This means that for this rule (only), the values of those two factors are irrelevant.

In other words, Rule 1 for Determine-Worth is equivalent to the rule:

```
IF PAYOFF/COST IS OVER 3
   THEN WORTH IS HIGH(100)
```

The (100) following HIGH means that for this rule I am 100% sure that the WORTH would be HIGH. In some other rules I might have multiple values listed with different levels of confidence in each value.

The listing of all factors in its rule base is an idiosyncracy of TIMM; many other shells simply list the relevant factors in each rule.

Another unique characteristic of TIMM is what's referred to as an (optional) secondary knowledge base for each mini-expert system, in which are listed combinations of factors that are impossible in the real world as indicated by the "#" value for EXPERTISE.

For KBE, I have included only one such secondary rule, which is in the Determine-Expertise mini-system:

```
Rule 1S
   If:
      NO. OF EXPERTS          IS OVER 3
      EXPERT AVAILABILITY     IS 50%-75%
      TURNOVER                IS LOW
      EXPERT ATTITUDE         IS INCOMPETENT
   Then:
      EXPERTISE               IS #
```

Whether or not it is a correct rule, this says that it is not possible to have OVER 3 experts who are available only 50%-75% of the time in a situation where the turnover is LOW and the experts' attitude is INCOMPETENT.

In another case I might have made the rule that it's impossible to have an application which is VAGUEly defined and also LOW risk.

Each expert system shell has its own ways of trying to simplify the decision problem, and this is one of TIMM's ways. By eliminating possibly large classes of solutions up front, the inference process may be simplified considerably.

KNOWLEDGE BASE FOR THE EXPERT SYSTEM "DETERMINE WORTH"

```
Rule 1
   If:
      PAYOFF/COST         IS OVER 3
      PERCENT SOLUTION    IS *
      TYPE                IS *
   Then:
      WORTH               IS HIGH(100)

Rule 2
   If:
      PAYOFF/COST         IS UNDER 1
      PERCENT SOLUTION    IS *
      TYPE                IS USEFUL
   Then:
      WORTH               IS NEGATIVE(100)

Rule 3
   If:
      PAYOFF/COST         IS 1-1.5
      PERCENT SOLUTION    IS *
      TYPE                IS USEFUL
   Then:
      WORTH               IS LOW(100)

Rule 5
   If:
      PAYOFF/COST         IS 1.5-3
      PERCENT SOLUTION    IS OVER 90%
      TYPE                IS CRUCIAL
   Then:
      WORTH               IS HIGH(100)

Rule 6
   If:
      PAYOFF/COST         IS 1.5-3
      PERCENT SOLUTION    IS <=50%-75%
      TYPE                IS *
   Then:
      WORTH               IS MODERATE(100)
```

```
Rule 7
   If:
      PAYOFF/COST          IS UNDER 1
      PERCENT SOLUTION     IS OVER 90%
      TYPE                 IS DEMO
   Then:
      WORTH                IS MODERATE(100)

Rule 8
   If:
      PAYOFF/COST          IS UNDER 1
      PERCENT SOLUTION     IS UNDER 50%
      TYPE                 IS CRUCIAL
   Then:
      WORTH                IS LOW(100)

Rule 9
   If:
      PAYOFF/COST          IS UNDER 1
      PERCENT SOLUTION     IS UNDER 50%
      TYPE                 IS DEMO
   Then:
      WORTH                IS LOW(100)

Rule 10
   If:
      PAYOFF/COST          IS UNDER 1
      PERCENT SOLUTION     IS 75%-90%
      TYPE                 IS NECESSARY
   Then:
      WORTH                IS LOW(100)

Rule 11
   If:
      PAYOFF/COST          IS UNDER 1
      PERCENT SOLUTION     IS UNDER 50%
      TYPE                 IS NECESSARY
   Then:
      WORTH                IS NEGATIVE(100)

Rule 12
   If:
      PAYOFF/COST          IS >=1.5-3
      PERCENT SOLUTION     IS OVER 90%
      TYPE                 IS *
   Then:
      WORTH                IS HIGH(100)
```

```
Rule 13
   If:
      PAYOFF/COST          IS UNDER 1
      PERCENT SOLUTION     IS <=75%-90%
   TYPE                    IS DEMO
   Then:
      WORTH                IS LOW(100)
```

KNOWLEDGE BASE FOR THE EXPERT SYSTEM "DETERMINE COMPLEXITY"

```
Rule 1
   If:
      INTUITION/COMMON SENSE  IS OVER 50%
      TECHNOLOGY              IS *
      DECISION DEFINITION     IS FUZZY
      KNOWLEDGE DOMAIN        IS *
   Then:
      COMPLEXITY              IS HIGH(100)

Rule 2
   If:
      INTUITION/COMMON SENSE  IS UNDER 10%
      TECHNOLOGY              IS EXISTS
      DECISION DEFINITION     IS WELL-DEFINED
      KNOWLEDGE DOMAIN        IS NARROW
   Then:
      COMPLEXITY              IS LOW(100)

Rule 3
   If:
      INTUITION/COMMON SENSE  IS 10%-50%
      TECHNOLOGY              IS MODIFY
      DECISION DEFINITION     IS OK
      KNOWLEDGE DOMAIN        IS OK
   Then:
      COMPLEXITY              IS MODERATE(100)

Rule 4
   If:
      INTUITION/COMMON SENSE  IS UNDER 10%
      TECHNOLOGY              IS BUILD
      DECISION DEFINITION     IS FUZZY
      KNOWLEDGE DOMAIN        IS NARROW
   Then:
      COMPLEXITY              IS HIGH(100)
```

```
Rule 7
   If:
      INTUITION/COMMON SENSE  IS OVER 50%
      TECHNOLOGY              IS BUILD
      DECISION DEFINITION     IS OK
      KNOWLEDGE DOMAIN        IS ECLECTIC
   Then:
      COMPLEXITY              IS HIGH(100)

Rule 8
   If:
      INTUITION/COMMON SENSE  IS UNDER 10%
      TECHNOLOGY              IS EXISTS
      DECISION DEFINITION     IS FUZZY
      KNOWLEDGE DOMAIN        IS ECLECTIC
   Then:
      COMPLEXITY              IS MODERATE(100)

Rule 9
   If:
      INTUITION/COMMON SENSE  IS OVER 50%
      TECHNOLOGY              IS EXISTS
      DECISION DEFINITION     IS WELL-DEFINED
      KNOWLEDGE DOMAIN        IS ECLECTIC
   Then:
      COMPLEXITY              IS MODERATE(100)

Rule 10
   If:
      INTUITION/COMMON SENSE  IS UNDER 10%
      TECHNOLOGY              IS BUILD
      DECISION DEFINITION     IS FUZZY
      KNOWLEDGE DOMAIN        IS ECLECTIC
   Then:
      COMPLEXITY              IS HIGH(100)
```

KNOWLEDGE BASE FOR THE EXPERT SYSTEM "DETERMINE EXPERTISE"

```
Rule 1
   If:
      NO. OF EXPERTS          IS 0
      EXPERT AVAILABILITY     IS *
      TURNOVER                IS *
      EXPERT ATTITUDE         IS *
   Then:
      EXPERTISE               IS UNAVAILABLE(100)

Rule 2
   If:
      NO. OF EXPERTS          IS 1-3
      EXPERT AVAILABILITY     IS OVER 90%
      TURNOVER                IS *
      EXPERT ATTITUDE         IS WILLING
   Then:
      EXPERTISE               IS AVAILABLE(100)

Rule 3
   If:
      NO. OF EXPERTS          IS 1-3
      EXPERT AVAILABILITY     IS UNDER 50%
      TURNOVER                IS LOW
      EXPERT ATTITUDE         IS INCOMPETENT
   Then:
      EXPERTISE               IS UNAVAILABLE(100)

Rule 4
   If:
      NO. OF EXPERTS          IS OVER 3
      EXPERT AVAILABILITY     IS OVER 90%
      TURNOVER                IS HIGH
      EXPERT ATTITUDE         IS WILLING
   Then:
      EXPERTISE               IS AVAILABLE(100)
```

```
Rule 5
   If:
      NO. OF EXPERTS          IS OVER 3
      EXPERT AVAILABILITY     IS 50%-75%
      TURNOVER                IS HIGH
      EXPERT ATTITUDE         IS HOSTILE
   Then:
      EXPERTISE               IS OK(100)

Rule 6
   If:
      NO. OF EXPERTS          IS 1-3
      EXPERT AVAILABILITY     IS OVER 90%
      TURNOVER                IS LOW
      EXPERT ATTITUDE         IS HOSTILE
   Then:
      EXPERTISE               IS OK(100)

Rule 7
   If:
      NO. OF EXPERTS          IS 1-3
      EXPERT AVAILABILITY     IS 75%-90%
      TURNOVER                IS MODERATE
      EXPERT ATTITUDE         IS UNINTERESTED
   Then:
      EXPERTISE               IS AVAILABLE(100)

Rule 8
   If:
      NO. OF EXPERTS          IS 1-3
      EXPERT AVAILABILITY     IS UNDER 50%
      TURNOVER                IS HIGH
      EXPERT ATTITUDE         IS WILLING
   Then:
      EXPERTISE               IS AVAILABLE(100)

Rule 9
   If:
      NO. OF EXPERTS          IS OVER 3
      EXPERT AVAILABILITY     IS 50%-75%
      TURNOVER                IS LOW
      EXPERT ATTITUDE         IS WILLING
   Then:
      EXPERTISE               IS AVAILABLE(100)
```

```
Rule 1S
   If:
      NO. OF EXPERTS           IS OVER 3
      EXPERT AVAILABILITY      IS 50%-75%
      TURNOVER                 IS LOW
      EXPERT ATTITUDE          IS INCOMPETENT
   Then:
      EXPERTISE                IS #
```

****

TIMM (TM) The Intelligent Machine Model
Version 2.0

****

KNOWLEDGE BASE FOR THE EXPERT SYSTEM "ASSESS CONTROLS"

```
Rule 1
   If:
      DATA CONTROL          IS *
      PROCEDURE CONTROL     IS LOW
      PERFORMANCE METRIC    IS *
   Then:
      CONTROL               IS LOOSE(100)

Rule 2
   If:
      DATA CONTROL          IS LOW
      PROCEDURE CONTROL     IS *
      PERFORMANCE METRIC    IS *
   Then:
      CONTROL               IS LOOSE(100)

Rule 4
   If:
      DATA CONTROL          IS *
      PROCEDURE CONTROL     IS *
      PERFORMANCE METRIC    IS LOW
   Then:
      CONTROL               IS LOOSE(100)

Rule 6
   If:
      DATA CONTROL          IS HIGH
      PROCEDURE CONTROL     IS >=MODERATE
      PERFORMANCE METRIC    IS >=MODERATE
   Then:
      CONTROL               IS TIGHT(100)

Rule 7
   If:
      DATA CONTROL          IS MODERATE
      PROCEDURE CONTROL     IS >=MODERATE
      PERFORMANCE METRIC    IS >=MODERATE
   Then:
      CONTROL               IS OK(100)
```

KNOWLEDGE BASE FOR THE EXPERT SYSTEM "ASSESS RISK"

```
Rule 1
   If:
      COMPLEXITY IS HIGH
      EXPERTISE  IS *
      CONTROL    IS *
   Then:
      RISK       IS HIGH(100)

Rule 4
   If:
      COMPLEXITY IS LOW
      EXPERTISE  IS AVAILABLE
      CONTROL    IS TIGHT
   Then:
      RISK       IS LOW(100)

Rule 5
   If:
      COMPLEXITY IS MODERATE
      EXPERTISE  IS OK
      CONTROL    IS OK
   Then:
      RISK       IS MODERATE(100)

Rule 6
   If:
      COMPLEXITY IS LOW
      EXPERTISE  IS AVAILABLE
      CONTROL    IS >=OK
   Then:
      RISK       IS LOW(100)

Rule 7
   If:
      COMPLEXITY IS MODERATE
      EXPERTISE  IS OK
      CONTROL    IS >=OK
```

```
   Then:
      RISK          IS MODERATE(100)
```

<u>Rule 8</u>

```
   If:
      COMPLEXITY  IS LOW
      EXPERTISE   IS AVAILABLE
      CONTROL     IS LOOSE
   Then:
      RISK        IS HIGH(100)
```

<u>Rule 9</u>

```
   If:
      COMPLEXITY  IS <=MODERATE
      EXPERTISE   IS UNAVAILABLE
      CONTROL     IS *
   Then:
      RISK        IS HIGH(100)
```

KNOWLEDGE BASE FOR THE EXPERT SYSTEM "DETERMINE SUITABILITY"

```
Rule 1
   If:
      WORTH                   IS HIGH
      EMPLOYEE ACCEPTANCE     IS POSITIVE
      SOLUTION AVAILABLE      IS NONE
      EASIER SOLUTION         IS NONE
      TEACHABILITY            IS FREQUENT
      RISK                    IS LOW
   Then:
      SUITABILITY             IS GOOD(100)

Rule 2
   If:
      WORTH                   IS NEGATIVE
      EMPLOYEE ACCEPTANCE     IS *
      SOLUTION AVAILABLE      IS *
      EASIER SOLUTION         IS *
      TEACHABILITY            IS *
      RISK                    IS *
   Then:
      SUITABILITY             IS POOR(100)

Rule 3
   If:
      WORTH                   IS LOW
      EMPLOYEE ACCEPTANCE     IS *
      SOLUTION AVAILABLE      IS *
      EASIER SOLUTION         IS *
      TEACHABILITY            IS *
      RISK                    IS HIGH
   Then:
      SUITABILITY             IS POOR(100)

Rule 4
   If:
      WORTH                   IS MODERATE
      EMPLOYEE ACCEPTANCE     IS NEUTRAL
```

```
      SOLUTION AVAILABLE     IS ADEQUATE
      EASIER SOLUTION        IS COMPLETE
      TEACHABILITY           IS DIFFICULT
      RISK                   IS HIGH
   Then:
      SUITABILITY            IS POOR(100)

Rule 5
   If:
      WORTH                  IS LOW
      EMPLOYEE ACCEPTANCE    IS NEGATIVE
      SOLUTION AVAILABLE     IS NONE
      EASIER SOLUTION        IS PARTIAL
      TEACHABILITY           IS FREQUENT
      RISK                   IS LOW
   Then:
      SUITABILITY            IS POOR(100)

Rule 6
   If:
      WORTH                  IS HIGH
      EMPLOYEE ACCEPTANCE    IS NEGATIVE
      SOLUTION AVAILABLE     IS PARTIAL
      EASIER SOLUTION        IS NONE
      TEACHABILITY           IS DIFFICULT
      RISK                   IS MODERATE
   Then:
      SUITABILITY            IS FAIR(100)

Rule 7
   If:
      WORTH                  IS HIGH
      EMPLOYEE ACCEPTANCE    IS POSITIVE
      SOLUTION AVAILABLE     IS PARTIAL
      EASIER SOLUTION        IS COMPLETE
      TEACHABILITY           IS FREQUENT
      RISK                   IS HIGH
   Then:
      SUITABILITY            IS POOR(100)

Rule 8
   If:
      WORTH                  IS HIGH
      EMPLOYEE ACCEPTANCE    IS POSITIVE
      SOLUTION AVAILABLE     IS PARTIAL
      EASIER SOLUTION        IS PARTIAL
```

```
      TEACHABILITY           IS POSSIBLE
      RISK                   IS LOW
   Then:
      SUITABILITY            IS GOOD(100)
```

Rule 9

```
   If:
      WORTH                  IS LOW
      EMPLOYEE ACCEPTANCE    IS POSITIVE
      SOLUTION AVAILABLE     IS ADEQUATE
      EASIER SOLUTION        IS NONE
      TEACHABILITY           IS FREQUENT
      RISK                   IS LOW
   Then:
      SUITABILITY            IS FAIR(100)
```

Rule 10

```
   If:
      WORTH                  IS HIGH
      EMPLOYEE ACCEPTANCE    IS NEGATIVE
      SOLUTION AVAILABLE     IS PARTIAL
      EASIER SOLUTION        IS NONE
      TEACHABILITY           IS FREQUENT
      RISK                   IS HIGH
   Then:
      SUITABILITY            IS FAIR(100)
```

Rule 11

```
   If:
      WORTH                  IS LOW
      EMPLOYEE ACCEPTANCE    IS POSITIVE
      SOLUTION AVAILABLE     IS NONE
      EASIER SOLUTION        IS COMPLETE
      TEACHABILITY           IS DIFFICULT
      RISK                   IS MODERATE
   Then:
      SUITABILITY            IS POOR(100)
```

Rule 12

```
   If:
      WORTH                  IS LOW
      EMPLOYEE ACCEPTANCE    IS NEUTRAL
      SOLUTION AVAILABLE     IS ADEQUATE
      EASIER SOLUTION        IS COMPLETE
      TEACHABILITY           IS FREQUENT
      RISK                   IS LOW
```

```
  Then:
     SUITABILITY             IS FAIR(100)
```

Rule 13

```
  If:
     WORTH                   IS LOW
     EMPLOYEE ACCEPTANCE     IS NEUTRAL
     SOLUTION AVAILABLE      IS NONE
     EASIER SOLUTION         IS NONE
     TEACHABILITY            IS DIFFICULT
     RISK                    IS LOW
  Then:
     SUITABILITY             IS FAIR(100)
```

Rule 14

```
  If:
     WORTH                   IS MODERATE
     EMPLOYEE ACCEPTANCE     IS POSITIVE
     SOLUTION AVAILABLE      IS ADEQUATE
     EASIER SOLUTION         IS NONE
     TEACHABILITY            IS DIFFICULT
     RISK                    IS HIGH
  Then:
     SUITABILITY             IS POOR(100)
```

Rule 15

```
  If:
     WORTH                   IS HIGH
     EMPLOYEE ACCEPTANCE     IS NEGATIVE
     SOLUTION AVAILABLE      IS ADEQUATE
     EASIER SOLUTION         IS PARTIAL
     TEACHABILITY            IS FREQUENT
     RISK                    IS HIGH
  Then:
     SUITABILITY             IS POOR(100)
```

Rule 16

```
  If:
     WORTH                   IS HIGH
     EMPLOYEE ACCEPTANCE     IS NEGATIVE
     SOLUTION AVAILABLE      IS PARTIAL
     EASIER SOLUTION         IS COMPLETE
     TEACHABILITY            IS POSSIBLE
     RISK                    IS LOW
  Then:
     SUITABILITY             IS FAIR(100)
```

<u>Rule 17</u>

```
   If:
      WORTH                  IS MODERATE
      EMPLOYEE ACCEPTANCE    IS NEGATIVE
      SOLUTION AVAILABLE     IS NONE
      EASIER SOLUTION        IS PARTIAL
      TEACHABILITY           IS DIFFICULT
      RISK                   IS HIGH
   Then:
      SUITABILITY            IS FAIR(100)
```

<u>Rule 18</u>

```
   If:
      WORTH                  IS MODERATE
      EMPLOYEE ACCEPTANCE    IS NEUTRAL
      SOLUTION AVAILABLE     IS ADEQUATE
      EASIER SOLUTION        IS PARTIAL
      TEACHABILITY           IS DIFFICULT
      RISK                   IS LOW
   Then:
      SUITABILITY            IS POOR(100)
```

APPENDIX C
Knowledge Base Access Method micro-PROLOG Standard Syntax

This appendix contains standard micro-PROLOG code for all of the relations referenced in Chapter 8, Frames: Data, Knowledge, and More, plus a few other relations. I tested these on an IBM PC with 256K of RAM and dual floppies, and on a Morrow Pivot I with the same configuration. These relations should provide a basic working set of frame access routines.

The relations are grouped according to their major purpose, although some relations are used in more than one section.

There may be more elegant ways to accomplish some of these tasks. The trouble with elegance in computer software is that it tends to obscure readability and hence maintainability. I have chosen the path of readability.

RETRIEVAL

1. Low-level Slot access relation.

```
((slot-of X Y)
  (frame (Y|Z))
  (APPEND x (X|y) Z))
```

2. Low-level Type access relation.

```
((type-of X (Y Z))
  (fget x (Y Z))
  (APPEND y (X|z) x))
```

3. Basic retrieval relation for Frame, Slot, Type, or Value.

```
((fget X Y)
  (frame (Y|Z))
  (EQ X (Y|Z)))
((fget X (Y Z))
  (slot-of (x|y) Y)
```

```
  (EQ x Z)
  (EQ X (Z|y)))
((fget X (Y Z x))
  (type-of (y|z) (Y Z))
  (EQ y x)
  (EQ X (x|z)))
((fget X (Y Z x y))
  (fget (z|X) (Y Z x))
  (ON y X))
```

4. Retrieve Value, Default, or invoke If-Needed at one level.

```
((fgetd X (Y Z))
  (fget X (Y Z Value)))
((fgetd X (Y Z))
  (fget X (Y Z Default)))
((fgetd X (Y Z))
  (fget (x y|z) (Y Z If-Needed))
  (true-of y X (Y Z)))
```

5. Retrieve Value through AKO inheritance, otherwise Default, or invoke If-Needed.

```
((fgeti X (Y Z))
  (fget X (Y Z Value))
  (/))
((fgeti X (Y Z))
  (fget (x y) (Y AKO Value))
  (fgeti X (y Z))
  (/))
((fgeti X (Y Z))
  (fget X (Y Z Default))
  (/))
((fgeti X (Y Z))
  (fget (x y|z) (Y Z If-Needed))
  (true-of y X (Y Z))
  (/))
```

6. Retrieve Value, Default, or invoke If-Needed at each level before AKO inheritance.

```
((fgetiz X (Y Z))
  (fgetd X (Y Z))
  (/))
((fgetiz X (Y Z))
  (fget (x y) (Y AKO Value))
  (fgetiz X (y Z))
  (/))
```

7. Retrieve class.

This is used to test if Y has an AKO link to X, e.g. to test if Bob is Human:

```
confirm(Human fgetcl Bob)
```

or to find all the frames to which Bob is linked by an AKO slot:

```
find(x:x fgetcl Bob)
```

```
((fgetcl X Y)
 (fget (Value X) (Y AKO Value)))
((fgetcl X Y)
 (fget (Value Z) (Y AKO Value))
 (fgetcl X Z))
```

UPDATE

1. Basic addition of Frame, Slot, Type, or Value.

```
((fput X (Y Z x y))
  (/* ADD A NEW FRAME)
  (NOT frame (Y|z))
  (EQ X (Y (Z (x y))))
  (add (X frame)))
((fput X (Y Z x y))
  (/* ADD A NEW SLOT)
  (NOT fget z (Y Z))
  (fget X1 Y)
  (APPEND (Z) ((x y)) z)
  (APPEND X1 (z) X)
  (/)
  (freplace X1 X))
((fput X (Y Z x y))
  (/* ADD A NEW TYPE)
  (/* Does the type exist)
  (NOT fget z (Y Z x))
  (/* Replace the type)
  (fget X1 Y)
  (fget Y1 (Y Z))
  (APPEND Y1 ((x y)) Z1)
  (replace Y1 Z1 X1 X)
  (/)
  (/* Replace in data base)
  (freplace X1 X))
((fput X (Y Z x y))
  (/* ADD NEW DATA)
  (/* Does the item exist)
  (NOT fget z (Y Z x y))
  (/* Replace the data)
  (fget X1 Y)
  (fget Y1 (Y Z))
  (fget Z1 (Y Z x))
  (APPEND Z1 (y) x1)
  (replace Z1 x1 Y1 y1)
  (replace Y1 y1 X1 X)
  (/)
  (/* Replace in data base)
  (freplace X1 X))
```

2. Database update.

```
((freplace X Y)
  (delete (X frame))
  (add (Y frame)))
```

3. Replace items in a list.

```
((replace X Y Z x)
  (/* Replace X with Y in Z giving x)
  (APPEND y (X|z) Z)
  (APPEND y (Y|z) x)
  (/))
```

4. Update Value and invoke If-Added through AKO links.

```
((fputia X (Y Z x y))
  (fput X (Y Z x y))
  (fputia z (Y Z)))
((fputia X (Y Z))
  (fget (x y|z) (Y Z If-Added))
  (true-of y X (Y Z)))
((fputia X (Y Z))
  (fget (x y) (Y AKO Value))
  (fputia X (y Z)))
```

5. Modify frame by adding in a Value previously available only through an fgetiz access path.

```
((fputiz X (Y Z))
  (fgetiz (x y) (Y Z))
  (fput X (Y Z Value y)))
```

6. Modify frame by adding in a Value previously available only through an fgeti access path.

```
((fputi X (Y Z))
  (fgeti (x y) (Y Z))
  (fput X (Y Z Value y)))
```

7. Modify frame by adding in as a Value the result of If-Needed.

```
((fputd X (Y Z))
  (NOT fget x (Y Z Value))
  (NOT fget y (Y Z Default))
  (fget (z X1|Y1) (Y Z If-Needed))
  (true-of X1 Z1 (Y Z))
  (fput X (Y Z Value Z1)))
```

8. Ask, the user relation most likely to be used in an If-Needed Slot during execution of an expert system for which data is needed from the user. Also here could be database retrieval relations.

```
((Ask X (Y Z))
  (pp Please enter a value for the Z slot of the Y frame :)
  (r X))
```

9. A similar message to be printed through the true-of relation.

```
((which-template true-of (X Y (Z x)) (Y)
     (Please give a Value for the x Slot in the Z frame)))
```

DELETION

1. Basic deletion of Frame, Slot, Type, or Value.

```
((fdel (X|Y) X)
  (/* Check for existence of the frame)
  (frame (X|Y))
  (/)
  (/* Delete it from the data base)
  (delete ((X|Y) frame)))
((fdel (X|Y) (Z X))
  (/* Check existence of the slot)
  (fget (X|Y) (Z X))
  (/* Delete the frame if one-slot)
  (fget x Z)
  (has-length x 2)
  (delete (x frame)))
((fdel (X|Y) (Z X))
  (/* Check existence of the slot)
  (fget (X|Y) (Z X))
  (/* Delete the old slot)
  (fget x Z)
  (purge (X|Y) x y)
  (/* Replace in data base)
  (freplace x y))
((fdel (X|Y) (Z x X))
  (/* Check existence of the type)
  (fget (X|Y) (Z x X))
  (/* Delete the slot if one-type)
  (fget y (Z x))
  (has-length y 2)
  (fdel z (Z x)))
((fdel (X|Y) (Z x X))
  (/* Check existence of the type)
  (fget (X|Y) (Z x X))
  (/* Delete the old type)
  (fget y Z)
  (fget z (Z x))
  (purge (X|Y) z X1)
  (replace z X1 y Y1)
  (/* Replace in data base)
  (freplace y Y1))
((fdel (X|Y) (Z x y X))
  (/* Check existence of the data)
  (fget (X|Y) (Z x y X))
```

```
  (/* Delete the type if one-data)
  (EQ Y ())
  (fdel z (Z x y)))
((fdel X (Y Z x y))
  (/* Check existence of the data)
  (fget X (Y Z x y))
  (/* Delete the old data)
  (fget z Y)
  (fget Xl (Y Z))
  (fget Yl (Y Z x))
  (purge y Yl Zl)
  (replace Yl Zl Xl xl)
  (replace Xl xl z yl)
  (/* Replace in data base)
  (freplace z yl))
```

2. Delete a frame.

```
((fremove X Y)
  (delete ((Y|X) frame)))
```

3. Delete from a list.

```
((purge X Y Z)
  (/* Delete X from Y giving Z)
  (APPEND x (X|y) Y)
  (APPEND x y Z)
  (/))
```

MISCELLANEOUS

1. List a frame.

```
((flist X)
  (? ((fget Y X) (PP Y))))
```

2. Determine list length.

```
((has-length () 0))
((has-length (X|Y) Z)
  (has-length Y x)
  (SUM x 1 Z))
```

3. Isolate the front X elements of list Z in Y.

```
((front X Y Z)
  (APPEND Y x Z)
  (has-length Y X))
```

4. Find the greater of two numbers.

```
((greater-of X (X X)))
((greater-of X (X Y))
  (LESS Y X))
((greater-of X (Y X))
  (LESS Y X))
```

5. Execute the relation X using Y and Z as arguments.

```
((true-of X Y Z)
  (X Y Z))
```

GLOSSARY

access method—*See* knowledge base access method.

active value—*See* daemon.

AND-OR tree—A graphic technique for representing the logical structure of a decision in which the final decision is shown as the top node of the tree. Determining conditions for the decision are shown on branches below the decision and joined by AND or OR, depending on the logical combination required for truth of the decision. Each of these second-level branches are similarly factored into the conditions required for determining their truth.

AI—*See* artificial intelligence.

AI workstation—A combination of computer hardware and software oriented toward developing AI systems.

analogical inference—A reasoning technique using production rules, in which a decision is reached by comparing it with the most similar previous decision or training experience.

analyst—An individual whose primary function is to create a specification of user needs as a prelude to system design and construction, possibly including working prototypes of the system.

apprentice—A computer system whose primary task is to assist a professional expert as an intelligent human apprentice would.

artificial intelligence—A field of endeavor whose goal is to devise computational models of intelligent human behavior. It includes such sub-fields and natural language understanding, robotics, expert systems, and automatic programming, as well as intelligent tutoring, design, and manufacturing.

automatic programming—A field of AI whose goal is to produce systems that can transform very high-level descriptive specifications of a user's needs into production code.

backward chaining—*See* chaining.

certainty factor—A usually informal measure of the likelihood that the premises or conclusion of a production rule are true.

chaining—A reasoning technique using production rules in which the truth of the conclusion of one rule leads to another rule for which that conclusion is a premise (forward chaining), or in which the need to prove the truth of a premise of one rule leads us to another rule for which that premise is a conclusion (backward chaining).

compiler—A computer system that translates statements in a high-level programming language like PL/I into an object code which is very close to native machine language.

conflict resolution—The process of determining which of two or more production rules are to be used when each may be applicable to a set of circumstances.

conventional—In this book, systems, hardware, and methods that have been thought of as belonging to data processing.

data base—A collection of facts organized into groups of records, each group having a unique set of identifying characteristics.

data base management system—A set of routines that act as an interface between physical records in a database and a program's logical need for information. Minimally, a data base management system provides facilities to add, delete, and retrieve records.

data flow diagram—A graphic notation for indicating the partitioning of a large system into several mini-systems and showing the information interfaces among the mini-systems.

daemon—A procedure stored as part of a frame structure that is activated when certain conditions in the frame structure are true.

DBMS—*See* data base management system.

decision space—For a given set of factors, each of which can take on certain values, the total number of possible combinations of factor values.

deterministic—Refers to a rule or procedure whose premises and conclusions are known with certainty.

DFD—*See* data flow diagram.

domain—An area of study or activity. Refers to a usually narrow area that may be the subject of an expert system project.

domain constraint—Information peculiar to a domain that may reduce the decision space in that domain, thereby limiting the need for exhaustive search when making a decision.

domain expert—One who is probably better at performing in a domain than those who are not considered to be expert.

expert—*See* domain expert.

expert system—A computer system whose goal is to make decisions or plan as well or better than experts in a particular domain.

expert system shell—A computer system that can reason, but has no knowledge with which to reason.

forward chaining—*See* chaining.

fourth-generation language—A class of descriptive data file management systems, such as FOCUS, RAMIS, and EASYTRIEVE.

frame—An information structure for representing complex interrelationships among symbolic information.

heuristic—A rule of thumb, or guideline, that can be applied to making a decision when we're not sure which path to take; after applying the guideline, we may still not know if the correct path was taken.

IF-THEN rule—*See* production rule.

inference—The act of reaching a conclusion based on a set of logical rules.

inference engine—A computer program that knows how to reason.

interpreter—A computer system that executes high-level programming language statements without first translating them into object code.

KBAM—*See* knowledge base access method.

KBMS—*See* knowledge base management system.

KEE—*See* knowledge engineering environment.

knowledge acquisition—The act of collecting from a domain expert the rules and techniques used by the expert in making decisions or planning.

knowledge base—A collection of factual information that includes the relationships among the facts.

knowledge base access method—A set of routines for manipulating the physical records in a knowledge base.

knowledge base management system—A computer system whose job is to act as an interface between the physical records in a knowledge base and a program's logical need for knowledge.

knowledge-based—Refers to systems in which knowledge is processed as a separate entity.

knowledge engineering—An activity similar to structured analysis, but including use of the techniques of knowledge-based systems as well as knowledge acquisition.

knowledge engineering environment—A software environment oriented toward the easy development of knowledge-based applications.

knowledge peripheral—A LISP machine or other knowledge-oriented processor used as a peripheral device to a mainframe computer.

LISP—A programming language which emphasizes the processing of lists of symbolic entities. The standard programming language for AI applications in the United States.

LISP machine—*See* AI workstation.

machine intelligence—*See* artificial intelligence.

mini-spec(ification)—The specification of a mini-system process that appears as a partitioning on a data flow diagram.

mini-system—A partition of a larger system represented as a circle on a data flow diagram.

natural English—*See* natural language understanding.

natural language understanding—The sub-field of AI whose goal is to allow fluent communication between user and machine in the user's conversational language.

non-deterministic—Opposite of deterministic.

partition—To factor a large system into smaller mini-systems.

performance metric—A way of determining the level of expertise at which an expert system is performing.

production rule—A primary way of representing knowledge for use in an expert system. The production rule has one or more conditions (the IF part) followed by a conclusion (the THEN part) that is true if the premises are true.

PROLOG—A descriptive programming language that implements the subset of logical statements known as definite clauses.

rapid prototype—A working model of a system. An early model constructed from the system specification.

robotics—A sub-field of AI whose goal is to build devices that can manipulate their physical surroundings in an intelligent way as a human does.

rule—*See* production rule.

rule base—A collection of production rules that form the knowledge base for an expert system.

shell—*See* expert system shell.

structured analysis—A discipline for developing a system specification of user needs that emphasizes data flow diagrams.

workstation—*See* AI workstation.

4GL—*See* fourth-generation language.

BIBLIOGRAPHY

Buchanan, BG & Shortliffe, EH, *Rule-Based Expert Systems,* Addison-Wesley, Reading, Massachusetts, 1984.

Charniak, E, Riesbeck, CK, & McDermott, DV, *Artificial Intelligence Programming,* Lawrence Erlbaum Associates, Hillsdale, New Jersey, 1980.

Clark, KL & McCabe, FG, *micro-PROLOG: Programming in Logic,* Prentice-Hall International, Englewood Cliffs, New Jersey, 1984.

Cramer, H., *The Elements of Probability Theory,* John Wiley & Sons, New York, 1955.

Davis, R & King, J, "An Overview of Production Systems," in Elcock & Michie (Ed), *Machine Intelligence 8,* Ellis Horwood, Chichester, England, 1977.

Dyer, MG, *In-Depth Understanding,* Yale University, New Haven, Connecticut, Research Report #219, May 1982.

Jung, CG, "Psychological Types," in *Collected Works of C.G. Jung,* Princeton University Press-Bollingen Series, Princeton, New Jersey, 1960.

Keller, R, *The Practice of Structured Analysis,* Yourdon Press, New York, 1983.

Kornell, J. "Embedded Knowledge Acquisition to Simplify Expert System Development," in *Applied Artificial Intelligence Reporter,* University of Miami, Coral Gables, Florida, Vol 1, No. 11/12, Aug-Sept 1984, pp28-30.

McCorduck, Pamela, *Machines Who Think,* WH Freeman Co., San Francisco, 1979.

Minsky, M., *A Framework for Representing Knowledge,* MIT AI Laboratory, Memo No. 306, June 1974.

Page-Jones, M., *The Practical Guide to Structured Systems Design,* Yourdon Press, New York, 1980.

Smith, RG, "On the Development of Commercial Expert Systems," in *The AI Magazine,* Fall 1984, AAAI, Menlo Park, California, pg 65.

Winston, PH & Horn, BKP, *LISP,* 2nd Ed., Addison-Wesley, Reading, Massachusetts, 1984.

Yourdon, E. & Constantine, L., *Structured Design,* Yourdon Press, New York, 1978.

Zionts, S., *Linear and Integer Programming,* Prentice-Hall, Englewood Cliffs, New Jersey, 1973.

Index